EXTREMELY BORI

NINE LIVES

"If you are a fighter pilot, you have to be willing to take risks."

Brig Gen Robin Olds USAF, Officer Commanding No. 1 (Fighter) Squadron RAF
February–October 1949

NINE LIVES

THE COMPELLING MEMOIR OF A COLD WAR HARRIER PILOT

CHRIS BURWELL

GRUB STREET | LONDON

AUTHOR'S NOTE

This is a personal account of my career in aviation which spanned the years 1969–2018. The views expressed are purely my own and are not necessarily the views of the companies or organisations for which I was working at the time.

This account is dedicated to my very good friend, Paul Hopkins, one of the outstanding pilots of my generation yet an exceptionally modest man.

Published by
Grub Street
4 Rainham Close
London SW11 6SS

Copyright © Grub Street 2022

Copyright text © Chris Burwell 2022

A CIP record for this title is available from the British library

ISBN-13: 978-1-911667-28-5

All rights reserved. No part of this publication may be reproduced, stored in a retrieval system, or transmitted in any form or by any means electronic, mechanical, photocopying, recording or otherwise, without the prior permission of the copyright owner.

Design by Myriam Bell Design, UK

Printed and bound by Finidr, Czech Republic

CONTENTS

FOREWORD

This book is the personal story of a highly professional pilot and very senior RAF officer. It captures not just the essence of Chris Burwell's flying in the service but also of his leadership and managerial experiences (both during and after his time in the RAF). Inevitably – and rightly – he has included some of the frustrations that confront all of us at various stages in our careers.

To those of us who have been through the RAF pilot training system, Chris Burwell's descriptions of his experiences (through officer, basic and then fast-jet flying training, and then on to his front-line service) will bring back many personal memories – both happy and tragic. Indeed, reading *Nine Lives* triggered me to look back through my flying logbook in order to remind myself of just how lucky I was to have been 'brought up' in the training system of the 1970s; my very earliest flying instructors were all in the RAF, mostly ex-fast jet, and they instilled in me (and many others) the confidence, skills and zest for life that carried us through to the front line. For Chris, that process saw him initially being 'creamed off' to be a first-tourist qualified flying instructor (itself a major accolade, and during which we first met), prior to a posting to the front line with the Harrier Force. I was then a young (and probably very naive but enormously enthusiastic) first tourist who, in serving on the same squadron as Chris, found a mature, wise and thoroughly trustworthy friend, both on the ground and in the air.

As for our shared experiences, the Harrier Force – particularly during the Cold War – was an exciting but very harsh environment. As Chris has said, that Force took the very top of the RAF's flying training output which resulted (I am ashamed to say) in an undeniable arrogance amongst fellow 'Harrier mates'. Jokes abounded about us: "Question: How can you tell if someone is a Harrier pilot? Answer: you don't have to as he will tell you himself within 30 seconds", or "Question: What is the difference between God and a Harrier pilot? Answer: God doesn't think he is a Harrier pilot". There was a highly uncomfortable truth in such quips against us but we didn't much care – we were a happy and fulfilled band of brothers, and Chris and I (and so many close friends) were part of that. We did, indeed, lose many colleagues (mostly in accidents and also on operations) but that was somehow accepted as being the price for taking part in the most exciting and demanding of all flying environments – low-level, single-seat fighter operations often from dispersed field sites, with the added challenge of VSTOL (during which any inattention, even for a second or so, could be fatal). The photograph of Chris and me socialising 'in the field' does, I hope, capture that deep camaraderie (see opposite).

Like Chris, I never did drop any weapon 'in anger' but we (by which I mean the UK's armed forces, in concert with all those of our NATO allies) had the greatest strategic victory of all – we 'won' the Cold War and thus helped to secure democracy in free countries in future years. It seems, however, strange to write such prose now, for the world is a more threatening place, so often typified by autocratic leaders whose very aim seems to be to destabilise all that we know and hold so dear. We must hope that leaders in the free world stand up to such threats.

But I must finish by confessing that there is one other reason why I jumped at the chance to write this foreword for, in all likelihood, I owe my life to Chris. To explain, when we were carrying out operations in Belize, he and I went 'up country' one weekend and, whilst swimming in the Rio Grande, I somehow contrived to slip on some very dangerous rocks above a waterfall and ended up face down, unconscious and bleeding, in the rocky shallows some ten feet below. Had Chris not been there to run around the waterfall and drag me out to safety, I would not be writing this now. So my thanks to him go far beyond gratitude for this intriguing book – and always will. There were very few people in the Harrier Force for whom I had a higher regard than he and I look back on those shared times with enormous affection.

Air Chief Marshal Sir Clive Loader KCB OBE RAF (Retd)
January 2022

INTRODUCTION

In 2006, my wife Lynne and I spent a weekend in Cambridge. While there we visited the impressive World War II American Cemetery and Memorial at Madingley where we fell into conversation with another visitor. He was an aviation enthusiast and when I told him about my background, he suggested I should write an account of my flying career. Pointing out that, unlike many of those buried at Madingley, I had done nothing exceptional as a pilot, he said that little had been written about flying for the military in more recent times and he was sure it would be of interest to aviation-minded people in the years to come. This suggestion germinated and, once I had retired (for the second time), I decided that I would write an account of my career in aviation, not least as a record for my three sons, of what I have done in my working life.

Serving in the RAF until I was 46 years old, reaching a reasonably senior rank level, and then moving on to work in two different aviation environments was not a normal progression. Whilst most military pilots looking for a second career in aviation go into airline flying, I went to work in a very different area – Aerial Work – and subsequently I spent four years in commercial flying training. I flew as a military pilot and as a civilian pilot and I was also a leader/manager in all the three areas of aviation in which I worked. When I came to making some notes about what to include in this account, I became more aware of something I knew already: that I had been very fortunate to survive my military flying career, thus the title of *Nine Lives*.

I would like to thank my erstwhile wingman and good friend Clive Loader for his generous foreword, and my subject matter experts – Ashley Stevenson, Mike Harwood, Mike Beech, Chris Rayner, Bob Marston, Gerry Humphreys and

'Socialising in the field'. The author and Flying Officer Clive Loader on loan to IV(AC) Squadron for TACEVAL at Eberhardt Harrier field site, Sennelager Ranges, West Germany in October 1978.

Wayne Morgan – for their help with making sure I got the facts right and, in some cases, with photographs. Huge thanks to my brother Dave for some pictures from his very extensive family archive and for his skilful editing of some very poor photographs. My thanks also to the team at Grub Street for their encouragement and deft editing; any errors or omissions are solely my responsibility. And finally, my heartfelt thanks to my wife Lynne for her encouragement of my 'vanity project' and for her ever perceptive comments, sage advice on what to leave out and tireless proof reading.

I cannot claim perfect recall and I apologise for any facts or points of detail that are not 100 per cent correct. I have tried to be honest and open in what I have written, at the risk of embarrassing myself on too many occasions, and the views expressed are mine alone.

Chris Burwell
Westwood
November 2021

CHAPTER I
EARLY DAYS

My earliest experience of flight was in the mid-1950s when I was four or five years old. I was on a family holiday to Pwllheli on the Llŷn Peninsula in Wales with my parents and my two older brothers when we visited the airfield at Caernarvon and flew in a Dragon Rapide. It was a warm summer's day and I recall the pilot in the front, the sun on his white shirt, and the particular aeroplane smell that would become familiar in the years ahead. Produced in the 1930s as a light commercial airliner, the Dragon Rapide, a twin-engined biplane, was an iconic aircraft of its time but, of course, this meant nothing to me as I watched the earth slip away to reveal an amazing view of green fields, the shoreline and distant mountains of Snowdonia from the skies of North Wales. Little did I think that in just 15 years' time I would be flying solo in a Gnat jet trainer over this same bit of countryside.

Like many children growing up since the earliest days of manned flight, I had a great interest in aircraft and was close enough to World War II to be aware of the vital role undertaken by the Royal Air Force during those dark days; there was no doubt

De Havilland Dragon Rapide, the first aircraft that the author flew in. (Keith Wilson/SFB Photographic)

Grasshopper glider in action at Uppingham School. (Uppingham School, Rutland)

that Bader, Gibson, Lacy, Cheshire and many others were heroes of my generation. Consequently, joining the RAF Section of the Combined Cadet Force at school was a natural choice when the opportunity arose in my early teens. This presented the opportunity to fly in Chipmunks at RAF Newton and RAF Wittering, including an early taste of aerobatics and, through a friend at school, a flight in a Canberra at RAF Bassingbourn which also involved lunch in the officers' mess, a foretaste of life to come.

In the Easter holidays of 1967, I went to RAF Swanton Morley in Norfolk on a one-week gliding course. This was carried out in the Slingsby Kirby Cadet Mark 3, a very basic tandem-seat aircraft dating back to 1935. Once we had learnt the basics of flying the aircraft – speed control, turns, circuit pattern, take-off and landing – we then had to master what to do if the cable launch failed, or the cable broke, during the take-off phase. This was all very exciting: depending on height, it required a very quick decision as to landing straight ahead when low, carrying out an abbreviated circuit if high enough, or carrying out an S-turn in-between to lose height before landing straight ahead. After a minimum of 20 dual flights, those who were assessed as sufficiently competent were dispatched on three solos. Undoubtedly this week, as much as any other experience, shaped the course of my future life.

Once back at school, my good friend Murray Inglis, who had been on the gliding course with me, and I were well placed to 'fly' the RAF CCF Section's Slingsby Grasshopper glider. This was an amazing contraption, based on a pre-WWII German design but dating back to 1952 in RAF service, comprising a fuselage frame (literally a two-dimensional frame) with a seat strapped onto it, wings and a conventional tail assembly. It was launched by three teams of helpers: two pulling on each end of a V-shaped elastic rope and one team restraining the glider at the tail until it was released by the pilot. Once the elastic rope was fully extended, the pilot released the restraint at the rear by pulling a toggle under the seat. The glider then accelerated across the grass before getting airborne for a very short 'hop', normally getting no more than about 30 feet into the air if that. If the grass was a bit long or wet, the additional resistance to the acceleration would result in the Grasshopper never actually leaving the ground! On the CCF Inspection Day in 1968 conducted by Air Commodore Evans, the officer in charge of operating the glider either set the controls incorrectly or failed to brief the pilot correctly with the result that the aircraft got airborne but went straight up into a classic stall. As a result, the nose dropped sharply and the glider impacted the ground heavily. The Uppingham School magazine recorded the event as follows:

> 'Air Commodore Evans was impressed with the wide range of training that he saw even if one cadet, finding himself 30 ft in the air in the RAF glider, could think of nothing better to do than plunge it straight into the ground.'

The wings folded forwards and the poor glider ended up as a sad pile of scrap wood and wire; fortunately, the pilot emerged from the wreckage unscathed. That was the end of our Grasshopper gliding at school.

During the 1960s my father learnt to fly and gained a Private Pilot's Licence (PPL) and I was therefore privileged to access some flying in light aircraft through his exploits at the Yorkshire Light Aircraft Club at Yeadon Airport (now Leeds Bradford Airport). Some of this was in a lovely old Auster V (G-APAF) which I never flew myself although my brother Dave learnt to fly on this aircraft. Given its tail wheel undercarriage configuration and its consequent propensity to ground loop[1] on landing, this was no mean feat. In 1967 I went with my parents to the USA for the first time on a charter flight in a Britannia arranged by the flying club at Leeds. This was a memorable experience, not least for the outbound flight which necessitated refuelling in Shannon to cope with the winds en route to New York. Whilst in America, we went to stay with a medical colleague of my father's who lived in Michigan and owned his own plane in which we flew up to the Great Lakes on a couple of days.

By this time I had come to the conclusion that flying was what I wanted to do in life; the question was whether it would be commercial or military. At the time I had

Kirby Cadet Mk. 3 as flown by many air cadets. (Keith Wilson/SFB Photographic)

a book about commercial flying which painted a very glamorous picture of life as a BOAC or BEA pilot and included the staggering fact that a captain could earn £5,500 a year. This just seemed too good to be true! However, I then saw an advertisement for RAF pilots showing two officers in their mess kit standing outside an officers' mess watching a fighter taking off in reheat at dusk. That clinched it. Getting rich would have to wait (or be abandoned), there was obviously much more exciting flying to be done in the military. Never showing any particular academic ability, my O levels had nevertheless gone better than expected and the two A levels I needed to get to the RAF College Cranwell to train as an officer and a pilot appeared to be achievable. I therefore made my application and was invited to the famous WWII fighter base at Biggin Hill for pilot selection.

Looking back at my visit to Biggin Hill, I do not think that I ever saw failure in selection as a possibility; I would like to think that this was not the arrogance of overconfidence, just the naivety of a 16-year-old boy but perhaps I am fooling myself. The whole process took four days although a number of candidates were discounted halfway through and were asked to pack their bags and return home. I was one of the fortunate ones and remained for the whole process which included a stringent medical (including balancing on one leg with your eyes closed); two interviews (one with two serving officers, the other with a 'headmaster' to determine your academic capabilities and potential); group exercises to assess your abilities as a team member and as a leader; giving a presentation on a topic of your choice (I chose potholing, something I had got involved in at school); and group discussions. I remember it as a very enjoyable experience and we certainly had some amusing times doing the group exercises in the hangar crossing the fabled shark-infested river with two planks, an oil drum and some rope. At the end of the week, we returned home without any idea of how we had fared. However, I was very pleased to receive a letter shortly afterwards to inform me that I was being offered a place at Royal Air Force College Cranwell (subject to my A level results), an RAF scholarship for my last year at school and a flying scholarship (30 hours flying on a light aircraft towards the issue of a PPL) which I could undertake the following summer. The only outstanding issue was that the medical had identified that my nose needed attention as I had a 'deviated septum'. This was the result of a game of chase in the swimming pool when my nose was the first part of me that came into contact with the bottom of the pool as I dived into the shallow end with my good friend John Young in hot pursuit. Following a minor operation, I was declared medically fit (A1G1Z1 in military parlance – A for air, G for ground service and Z for overseas service) and ready to commence training in September 1969. But, before then I had to acquire two A levels, which I managed (just), and had to report to Air Services Training in Perth, Scotland, for a month to complete my flying scholarship. Below are my thoughts on that day; written years later:

GOING SOLO – JULY 1969

Four weeks ago, I finished school. Today I am taking my first real steps as an adult and, to prove the point, unusually, both my parents have come to see me off at the railway station. Perhaps their behaviour is born of concern: their third and youngest son, still 17, and today I am going to Scotland for the next month to learn to fly. Such parental anxiety is lost on me in the sheer excitement of starting out on what I hope will be a career in flying. Even the train journey northwards is an adventure, a solo step into uncharted territory. The RAF has awarded me a flying scholarship and they have sent me to Scone airfield near Perth where I will be given 30 hours of flying instruction. Although most of us going to Perth for training will fly the Cessna 150, two of us will learn on the Chipmunk. The de Havilland Chipmunk is a venerable aircraft, a derivative of the world-famous Tiger Moth, and I want to fly it because it has a tail wheel (making it more demanding to handle on the ground), but particularly because it is capable of carrying out aerobatics. I have written to Air Service Training at Perth and I have asked them if I can fly this aircraft. But tonight as we gather at this remote airfield in rural Perthshire and get to know each other, there is a more pressing subject: the small matter of Armstrong and Aldrin landing on the surface of the moon. Michael Collins is orbiting the moon alone and, if all goes well, we will be going solo very soon as well. Tomorrow we start on our ground school instruction: Rules of the Air, Principles of Flight, Meteorology, Navigation and Aircraft Technicals.

Some two weeks later I am going out to fly some more circuits in the Chipmunk with my instructor, Captain Gordon Lockhart. It is a fine summer's day with scattered cumulus cloud, a light breeze down the runway and excellent visibility. I go out to the aircraft on my own to carry out the pre-flight inspection: chocks in position, fire extinguisher to hand, an initial check of the switches in the cockpit followed by a careful walk round of the aircraft. I am looking for any sort of damage to the airframe, I must ensure the tyres are in good condition, the pitot tube is clear; I prime the engine with fuel and check the security of the canopy escape panel. Once complete, I can now climb into the cockpit, strap myself in, and carry out the 'left to right' cockpit checks in preparation for starting the engine. Whilst doing all this, Gordon Lockhart climbs into the back. After a check that the intercom is working, we are ready to start. I check the fuel is on, the brakes are on, throttle set half an inch open, magneto switches on, and pull the starter. There is a very loud bang and a wonderful smell of cordite. This aircraft has a cartridge starter which means that a shotgun cartridge is used to turn the engine for starting. The propeller turns over, and with a cough and a splutter the engine leaps into vibrating, shaking life. I check the oil pressure is rising, set 1200 RPM and check the generator light is out, then carry out the after-start checks. We get taxi clearance and slowly snake our way towards the runway, turning continuously, as I cannot see what is in front of the aircraft due to the high nose in front of me.

I am overjoyed. I am doing what I have wanted to do for some years even though it is demanding. I am confident in what I am being asked to do today but I know now that this aircraft can make demands of me that I cannot yet meet. A few days ago, Gordon Lockhart introduced me to aerobatics. We dived down until we were flying at 120 knots then pulled straight up through the vertical, over the top upside down, then down the other side through the vertical and pulled out of the dive: a loop. This is where aerobatics start. And the sensation of the 'g' force, the visual impact of the sunlit Scottish countryside being turned upside down and the sudden discovery of mastery over this element is an incredible revelation. (That loop used 800 feet of sky; in seven years' time I will fly my first loop in a Harrier and that will use 7,000 feet of sky). But having now done some aerobatics with Gordon Lockhart, I am aware of the challenge of having the confidence to carry out aerobatics on my own.

Pre-take off checks complete, the control tower clears us for take-off. Carefully I point the aircraft straight down the runway and slowly open the throttle. Learning to fly on the Chipmunk is not recommended for those with a nervous or under-confident disposition because at this point, as the power takes effect, the aircraft tries to swing hard towards the side of the runway. A positive 'boot full' of rudder is required to counter this. As the speed builds and I ease forward on the control column to lift the tail of the aircraft off the runway, the Chipmunk produces its next trick: it now tries to swing in the opposite direction. A boot full of rudder in the other direction tackles this effect and with the speed now coming up to 70 knots I ease gently back on the control column to get airborne. I put the nose into the climbing attitude, trim the aircraft[2] and settle down to climb to 500 feet, then turn left through 90 degrees onto the 'crosswind leg'. Once level at 1,000 feet, I allow the aircraft to accelerate to 90 knots, reduce the power, re-trim, then turn left through a further 90 degrees to roll out on the 'downwind leg'. There is little time to do or think about anything other than flying the aircraft accurately and carrying out the drills I have been taught.

The aircraft I am flying has the registration G-APLO, or Lima Oscar as she is known on the radio and to those who fly her. There are two Chipmunks at Perth but I have done most of my flying on Lima Oscar and already I feel an affinity for this neat, beautifully maintained and delightful aircraft. Gordon Lockhart has made it abundantly clear that I should always handle her delicately; if I treat her right, she will respond and will perform correctly. (I will never forget my first aircraft and in 36 years' time I will be sharing the same airspace and radio frequency with her over the Channel Islands, the first time I have seen her since Perth. I am flying a Beech King Air twin turboprop aircraft. On the day I went off into the circuit with Gordon Lockhart I had ten hours flying experience; sitting in the King Air I had over 6,500 flying hours. But I had never forgotten Lima Oscar. Abandoning radio discipline, I announce on the radio that Lima Oscar was my first solo 36 years ago and her pilot assures me that she is still doing well. We pass in the air 1,000 feet apart and she looks immaculate.)

But back to that momentous day. I land and take off a number of times under Gordon Lockhart's scrutiny and then he asks me to taxi clear of the runway. Without further ado, he tells me I am now on my own, to carry out a circuit and landing and taxi back in. He climbs out of the aircraft and walks away across the grass.

CHAPTER 2
RAF TRAINING 1969–72

On the afternoon of 28 September 1969, 60 aspiring officers arrived at Newark railway station in Lincolnshire to join No. 100 Flight Cadet Entry at the Royal Air Force College Cranwell. We were quickly shepherded onto the waiting RAF coaches by Sergeant Les Rodda who, I was about to discover, was to be my senior non-commissioned officer on B Squadron for the coming six months. On the bus I sat next to Don Bishop who was also assigned to B Squadron and was in the next room to me in B Squadron's junior mess block. I got on well with Don on the journey to Cranwell, despite the fact that we were both undoubtedly filled with trepidation at the unknowns ahead of us, and he proved to be a good friend throughout our two-and-a-half years' training at Cranwell.

At the time I joined Cranwell, the fight cadet system of bringing trainees into the RAF straight from school was being run down in favour of the graduate system, with the majority of future officers coming from university. Whilst flight cadet entries had joined at six-month intervals, we arrived one year after 99 Entry, and 101 Entry, the last-ever cadet entry arrived one year after us. Each entry was made up of aspiring pilots, navigators, engineers, administrators, supply and RAF Regiment officers. With the changeover to the graduate system, on arrival at Cranwell we were given the choice of trying to gain a university place through university clearing and going straight off to university now as acting pilot officers; staying at Cranwell for one year, retaking A levels then going to university; or staying at Cranwell for the full two-and-a-half-year course before being commissioned as pilot officers and, in the case of aircrew, being awarded the flying brevet at the same time. A limited number of people did opt for one of the university routes but the majority, myself included, had our minds set on learning to fly and getting through training as soon as possible and elected to stay on at Cranwell for the full course, a decision I never regretted.

The Cranwell flight cadet year was divided into two terms. After kit issue and the first haircut (mandatory, regardless of how short your hair was already!) our first term was largely taken up with drill, physical exercise, sport, learning about the military and the RAF, the role of an officer, leadership, air force law, field craft (living in the field), RAF customs and mess etiquette including how to use a knife and fork. In the first week we were shown how to 'bull' our kit and prepare our rooms for inspection and we had mentors from 99 Entry to show us the ropes. I was fortunate to have Dave Payne, an ex-apprentice, who looked after us very well; sadly he was killed flying helicopters some years later. After a period of grace lasting little more than a week, we were subjected to 'crowing' in the evenings which involved having our rooms, kit and the communal ablutions inspected by members of 97 Entry (the senior entry) for a number of weeks. If they did not like what they found they would throw our belongings on the floor and tell us to start all over again or give us punishments or tasks to perform.

Throughout this period, we were not allowed off camp at all and could only visit the bar to buy soft drinks and there was certainly no time for idle socialising. All this helped mould us into a cohesive group of youngsters and the 'end of restrictions' party at the end of six weeks, once we had met the required standard, was something of a riot. Throughout these early weeks, we were all pretty tired most of the time with non-stop 18-hour days and lots of physical exercise. As an able cross-country runner, I was chosen to represent the college, which was good in that it gave me approval for weekends off camp, but it did mean I was now doing even more physical training. I had never been so fit and remember waking up in the middle of one night with the most awful cramp in my calf. I knew the only thing to do was get out of bed and exercise my leg so threw the covers off the bed and, still half asleep, pulled my knees up straight into my nose as I sat up to get out of bed. I finished up next to the bed in agony unsure whether it was my leg or my nose which hurt more.

Once we had mastered the basics of drill, we were required to form up at 0630 outside our accommodation block and march ourselves to the armoury to collect our rifles before reporting to the parade ground for drill at 0700. One morning as we marched to the armoury, my good friend Dave Monteith, who was directly in front of me broke ranks and ran off back towards the block. A little later, Dave caught up with us at the armoury. Asking him what the matter was, he said he had suddenly realised that his feet were not making the same noise as everyone else and when he looked down, he saw that he was still wearing his slippers instead of his drill boots.

We had an excellent bunch of senior non-commissioned officers (SNCOs) who trained us on the drill square throughout our time at Cranwell. Sergeant Rodda turned out be a lovely guy – shouty and demanding on the outside but with a light touch and an avuncular concern for our successful progression to becoming commissioned officers on the inside. Not necessarily one of the brightest of SNCOs, he had us in stitches one day with his order of 'right gentlemen, pair off in threes'. I felt really bad one morning when, quite unusually, I managed to sleep in and was late on parade for the weekly Wednesday morning drill practice with the whole college. This was a chargeable offence so when Sergeant Rodda asked why I was late, I came up with the excuse that I had had the 'squits' all night and had hardly had any sleep. He spent the whole of the drill session coming up behind me asking 'Are you all right, Mr Burwell?'. When I said I was doing OK, he would say 'Good man!' before returning ten minutes later to make sure I was still OK. The god of the drill square was Warrant Officer John Garbutt who was the epitome of what an RAF warrant officer should be: immaculate, sharp and totally focused. One of his best quotes was: "Gentlemen, I shall call you Sir and you will call me Sir. And the difference is, you will mean it!"

Previous flight cadet entries had had the challenge and motivational experience of flying the Chipmunk during their first term at the college. However, with the expansion of the university graduate scheme, the Chipmunks had been removed from Cranwell to bolster the university air squadrons. Consequently, we were not planned to fly until spring 1971. Whilst this was frustrating for us all, at least I had already done some flying, but a number of the pilot cadets on 100 Entry had never flown in anything other than an airliner and would have to wait for 18 months before they could fly an aircraft themselves for the first time. The only positive side to this was that we would start our flying straight onto jets, on the Jet Provost Mk. 3, skipping the usual piston training; this was an exciting prospect for us all.

In the Easter break in 1970, at the end of our first term, we were all required to spend a week visiting an RAF station. Malcolm Howell and I were sent to RAF Leuchars in Scotland where we spent each day being shown around various squadrons and sections of the station. My lasting memory of the visit was being ushered in to meet OC 43 Squadron. The squadron had recently re-equipped with the Phantom, which was new into RAF service. There was decoration going on in the squadron building and, as I entered the boss's office and offered my best flight cadet salute, I was not surprised to see large sheets of paper over much of the floor, presumably to protect the carpet from the paint. OC 43 Squadron stood up at his desk and offered his hand. As I walked smartly up to his desk across the paper on the floor and shook his hand, he looked me in the eye with the well-practised stare of a seasoned fighter pilot and senior officer and said, "Get off my f***ing maps, I need those to fly across the Atlantic tomorrow". Having embarrassed myself suitably in the real air force, it was something of a relief to return to term two at Cranwell.

The next year (terms two and three) were taken up with building on what we had learnt in term one and sitting through a lot of academic training; I certainly remember being subjected to 120 hours of thermodynamics and an extensive module on helicopter aerodynamics over this period. We also did lots more physical exercise, sport, field exercises and drill. The field exercises included camping and walking over the mountains of Scotland in the middle of winter and a very demanding weekend exercise in the Catterick Ranges in North Yorkshire in January (it was certainly incredibly cold):

ON ARKENGARTHDALE MOOR

Oh barren moor why do we cross
Your field less wastes and muddy mass
Chill in the wind that blows across
Your open acres?

Shelter's little that you afford
And day is brief when night is cold
Oh heartless home where I before
Have dwelt a night and slept within
Your icy jaws.

Cranwell 1971

Exercise King Rock in Germany was the main exercise of the whole course and involved training in a number of outdoor activities and ended with a four-day escape and evasion exercise based on a crew having parachuted from their aircraft over enemy territory. We were split into groups of three or four for this phase and were dropped off in the middle of the night from the back of a lorry with the back fastened down at an unknown location (apart from that we were somewhere on the map we were given). We had to make a rendezvous point (RV) some time the next night to get our next RV location and time and so on for three days and four nights. We had been informed that all the RAF staff on the exercise, plus a regiment of the British Army and the local police would be trying to capture us. It was self-evident that moving under cover of darkness was the only way to operate. However, it is very difficult for the body clock to accommodate this; when flying across time zones, the body will adjust at only two hours per day.

The first day we all fell asleep as soon as we had built a shelter but were awake again a couple of hours later and had to spend the next 12 hours killing time and getting very bored until we could move off at dusk. In addition, the only food we were given was aircrew survival rations which would give us enough calories to survive in a static location, not walk ten miles or more every night. Consequently, even by the second night, we were all showing serious signs of fatigue, most notably falling asleep as soon as we stopped for a rest during the night's walking. It was necessary to move quietly, speaking only in whispers, to avoid giving ourselves away; we also had to keep alert for signs of the 'enemy'.

Given fine weather and clear skies navigation was not too difficult but it could be very dark when the moon was not up. We had to avoid farm houses as much as possible as the dogs would hear us from a long way off and start barking which would let people know we were in the area. On the second night, we had to cross a valley. There were three bridges across the valley, one of which was the obvious one to use (and was therefore most likely being watched), whilst the other two would require lengthy detours, north or south. Given our physical and mental state by then, we elected to go for the easy option and having spent some time watching the bridge and its approaches for signs of the 'enemy', decided it was safe and started making our way across it. At this point, an army armoured personnel carrier (APC) came down the track behind us and turned onto the bridge. We were now just clearing the bridge on the far side and in our panic to get away two of us turned left and the other two to the right. The two of us threw ourselves into the undergrowth, covered ourselves with vegetation and lay absolutely still. The APC came off the bridge and turned left towards us and stopped very close. I could not see how close as I had my head buried in my beret. We heard someone say "we'll get a brew on then wait here as they're bound to be coming over the bridge very soon". The next thing I knew, I was waking up as I was so tired that I had fallen asleep, as had my colleague next to me. I nudged him awake and we very quietly walked away down the lane where we met up with the other two who were waiting for us. Thinking back, it was quite fortunate that the APC did not run us over or stop on top of us as we were very well camouflaged against the forest floor.

There were a couple of other amusing incidents during the nights we walked across the German countryside evading our would-be captors. On one occasion we were crossing a large open field and could hear horses in the field with us but could not see them. After a short time, there were sounds of galloping coming straight towards us (or so we imagined). At this point the four of us all ran in panic, in totally different directions, trying to get out of the field as quickly as possible to avoid being trampled underfoot. Having all escaped from the field at totally different points, it then took us about 30 minutes to get everyone back together again (as previously mentioned, we could not call out for fear of giving ourselves away so had to rely on stumbling around in the dark and calling out in whispers). On another occasion when it was again very dark, we were crossing a field but this one had a dip in the middle running right across it which appeared to be a stream with no easy way to cross it. After some discussion, I whispered, "OK, I'll jump and see if I can make it across". Perhaps unsurprisingly, everyone agreed that I should do this (and not them). I lunged forward, my shins came into contact with a low wire fence running down the length of the stream which we had not seen and I went face first into the water on the far side! Needless to say, the other three found this hugely amusing and the ensuing laughter would surely have given away our location if anyone had been around to hear us.

On the third day we were holed up in a small wood and built a four-man shelter as usual with our one parachute, the only tent/sleeping material we had with us. We were beginning to think that we might make it through to the end of the escape and evasion phase without being caught when some of the RAF staff appeared and marched us off into custody. In our run-down state we had not been thinking clearly and had left obvious tracks in the undergrowth which had led them straight to our shelter. In the event we were only retained for a limited time then sent off again to make the final RV that night which marked the end of the exercise. This phase was certainly a good lesson in the limitations of the human body and mind to cope with such a situation even without the added stresses of an actual aircraft abandonment, inclement weather and an enemy that would quite probably shoot to kill. I would think back to these experiences when I found myself flying over Northern Iraq some 22 years later.

There was a tragic occurrence early in term two: one of the cadets on 98 Entry flew into the ground at night and was killed. Although this was an accident occurring in basic flying training, by virtue of the high performance of many of the aircraft types in use in the military, and the roles being undertaken by all front-line aircraft, military aviation is inherently risky. This was our first exposure to the fact that the loss of friends and colleagues in flying accidents is (and always has been) a difficult feature of military aviation. We all had to learn to live with this if we were to make careers as military aviators.

My good friend Dave Harle and I had resurrected the College Potholing Club which had been dormant for some time. We both had some limited experience of caving and potholing but realised that pursuing this gave us a legitimate excuse for being out of the college at weekends. Consequently we built up a core of regular members who would join us on trips to North Yorkshire (staying at RAF Leeming) to get away from Cranwell and to discover what this ridiculous sport was all about. On one memorable occasion, we took a party including Laurie Barnes (initially a pilot but transferred to RAF Regiment and retired as a group captain) down Sunset Pot near Chapel-le Dale in the Yorkshire Dales. With a number of 'first timers' (including Laurie), before entering the cave I explained the operation of the carbide light and how to recharge it. (The carbide light, attached to the potholer's helmet, has calcium carbide in the lower chamber and water in the upper chamber. Water dripping onto the calcium carbide creates acetylene gas which is burnt, giving a bright flame which provides sufficient light to explore the cave.) Once we reached the final chamber in Sunset Pot, I then demonstrated how to recharge the light, emphasising the need to ensure that the rubber washer remained in place when you screwed the two chambers of the light back together again to avoid the loss of gas. When his turn came, Laurie went through the drill, replaced his helmet with the flame burning brightly, at which point there was a soft 'wooomph' as Laurie's complete light burst into flames on his head as he had missed out the washer and the escaping acetylene had ignited around his helmet.

We also had to call out Cave Rescue on one occasion. Dave Harle was leading a party (I was exploring another cave), when a stone dislodged by Roger Booth at the top of a ladder pitch hit Chris Gash waiting at the bottom of the ladder in the back. Initially it was thought that the injury was quite serious but, by the time Cave Rescue arrived, Dave had got everyone out of the cave and the worst of it was some serious bruising. The Cave Rescue team was very understanding and assured us that they were perfectly happy to come out as a precaution.

Eventually, in the spring of 1971 it was almost time for us to start our long-awaited flying training. However, before we could take to the air, we had to complete the ground school syllabus and pass the Command Examination Board examinations to qualify for the award of our pilot brevets. This involved many more weeks in the classroom studying a wide range of subjects such as aerodynamics, meteorology, air law, navigation, electrical and hydraulic systems and learning Morse code. Now, at last, we had the start of flying in our sights but this still couldn't happen until we had been kitted out with a flying suit, thermal underclothes, flying boots, flying gloves, cold weather jacket and a 'bone dome' (flying helmet complete with oxygen mask) and leg restraining garters to attach your lower legs to the ejection seat.

At last, on 7 April 1971, I flew my first RAF flight in a Jet Provost Mk. 3 with my instructor, Flight Lieutenant Brian Synott. Flying the Jet Provost involved getting dressed in all the right kit (somewhat in contrast to flying the Chipmunk at Perth), going through a detailed brief with your instructor, checking the Form 700 for the aircraft, carrying out a careful external check of the Jet Provost, then going through the process of strapping into the ejection seat with the help of the ground crew before you got anywhere near to starting the engine. My first impression of the 'JP' was that it was much smoother and easier to fly than the Chipmunk. You could actually see where you were going on the ground and, whilst it was obviously somewhat faster than the Chippie, with a single-jet engine, it didn't have any of those 'interesting' directional issues on take-off. Under the considerate and capable tutelage of Brian Synott I progressed happily through the early dual exercises and carried out my first jet solo at RAF Barkston Heath. In due course, my instructor changed to Flight Lieutenant Bob Jones and I clearly remember learning to fly a barrel roll with him one day out over the Trent valley. He was carrying out a demonstration and as he started pulling up and rolling, I looked up to see that we were pulling straight up into the well-camouflaged underside of a Vulcan bomber! We were so close that all I could do was yell "I have control" and roll to the inverted and pull down to avoid flying into the other aircraft. Bob was somewhat bemused as he never saw the Vulcan until we had recovered from our impromptu manoeuvre and turned round to see it flying away quite happily blissfully unaware of the near miss. It was somewhat ironic that Bob had just come to instructing from flying Vulcans.

Quite early on in our flying training, Paul Hopkins and myself were selected to be converted early to the more powerful and faster Jet Provost Mk. 4 and by mid-July I was completing my basic handling test and ready to move on to the applied exercises. In October, after completing night flying, I converted to the Jet Provost Mk. 5 (effectively a JP 4 with a pressurised cockpit) and started on the long-awaited formation phase of the course – now we were all Red Arrows pilots in the making! To be so close to another aircraft in the air was incredibly exciting if not a little scary at times. After learning to stay in close formation in straight and level flight, we then progressed to breaks and re-joins. During the re-joins we had to be careful to control the overtake speed, and to ensure we always stayed low on the leader, until we had stabilised two wing spans out to preclude the risk of collision. We then moved on to maintaining close formation during turns which were very gentle initially but then progressed to more bank and a little G force until, towards the end of the phase, we were carrying out big wingovers. From the start of formation training we were introduced to tail chasing where the leader flew various manoeuvres including aerobatics and we had to follow at about 200 yards separation. During this we learnt about lead and lag which would come in useful for those of us going onto fighters and would apply this to air combat tactics. After four dual formation flights we were sent off solo. I was paired up with Paul Hopkins, who turned out to be the outstanding pilot on my entry at Cranwell. One of the instructors would take Paul and I off for an hour, each of us solo on either side of him, carrying out the same profiles that we had done dual and working us hard. It was obvious that Paul was good; as the manoeuvring increased and I started to wobble around, I could see Paul on the far side sitting dead steady as though he had been born to it.

With night flying and formation completed, we had some more navigation flights to carry out including landaways, dual (Leeming and Leuchars) and solo (to Leeming), more instrument flying to gain a white instrument rating and eventually the final handling test, the last milestone to qualify for the award of the pilot's brevet. I completed this on the morning of 23 December 1971 just before we finished for Christmas at lunchtime. We were not due to graduate from the college until late February but in mid-January I was back in the air training for the Battle of Britain aerobatics competition which I won. Even to this day, I am convinced that I was put up for this competition ahead of Paul Hopkins to give someone else a chance to win at least one of the flying prizes (Paul did get the other two).

With everyone finishing their flying early in the New Year, we were sent off to RAF Abingdon to complete a parachuting course. This was not a part of the normal flight cadet course at Cranwell but, since Prince Charles had recently carried out a course, we

Newly commissioned pilot officer with Wings, Graduation Day RAF College Cranwell on 25 February 1972.

were all sent off to do two weeks to fill in the time before graduation. Contrary to the expectations of some, the course was great fun and, after a week of rolling around the hangar floor and jumping off various bits of apparatus, the dreadful weather in our second week that stopped us actually jumping was a great disappointment. Unfortunately, this training did not prevent me from damaging myself on my first and only parachute descent some seven years later! In late February we were back at Cranwell for our graduation and to get our postings. I had got my first choice and had been selected for the fighter stream – I was off to RAF Valley in Anglesey for my advanced flying training on the Gnat. On 24 February I was awarded the RAF pilot's brevet by the commandant, Air Vice-Marshal Fred Hughes[3], and the following day we were commissioned as pilot officers on the graduation parade.

Having enjoyed a few weeks' leave, I next found myself unexpectedly reporting for duty at the Transport Command base at RAF Brize Norton instead of Valley in Wales. I had received a message to go to Brize as the Gnat fleet had been grounded following a fatal crash when the tail assembly had come off one of the aircraft. We were told that we might have to wait six months before going to Valley. Joining 511 Squadron (Britannias) as a 'holding officer' (dogsbody), I was given various mundane tasks to carry out. It quickly became apparent that the best course of action to avoid these jobs would be to get assigned to a flight to somewhere interesting as supernumerary crew and make myself scarce. Needless to say (at least for those who know how the military works), as soon as I had got myself lined up for a trip to the Far East, a signal arrived ordering me to report to Valley the following week as the Gnat was now cleared to fly again.

In early April Paul and I, amongst others, arrived at Valley to join 67 Course. The sun was shining, the beaches looked fantastic and we were all going to be fighter pilots. There was only one small blot on the landscape: we had to undergo a number of weeks back in ground school before we could get our hands on a Gnat. In the event, ground school passed quickly even if I did have a lot of trouble getting my brain round high-speed aerodynamics. I also recall a briefing we had on oral hygiene from one of the station dentists which included the exhortation to keep our dental records up to date as this was often the only way to identify our bodies after we had killed ourselves flying.

Since the runways at Valley were undergoing resurfacing, we commenced our flying at RAF Fairford in June 1972. The Gnat was a fabulous aircraft; surprisingly small, it was said that you didn't climb into it but you pulled it on! With a much higher thrust-to-weight ratio than the JP and hydraulically powered flying controls, the acceleration on the runway was impressive and it was a fast and very manoeuvrable aircraft. Nevertheless, it did have a couple of vices that could catch you out. Firstly, the longitudinal control system could be a major problem in the event of a hydraulic failure (all Gnat pilots will recall the mnemonic STUPRECC[4] to help remember the emergency drill in the event of a hydraulic failure). Secondly, with its narrow track undercarriage, the aircraft could be difficult to control on the runway. This vice would be exacerbated hugely if the braking parachute were used on the landing run in a crosswind, with the aircraft doing its best to depart off the side of the runway.

The content of the course at Valley was similar to what we had done at Cranwell but in half the flying hours and twice the speed. I was fortunate to have Ross Payne, a Lightning pilot, as my flying instructor and I progressed happily through the course. Towards the end of the course, Ross went on leave and I flew with another instructor for that week. I did not enjoy flying with him, my performance suffered and I realised he was not happy to be instructing. By the end of the week, I was having something of a crisis of confidence in my ability. The following Monday when Ross returned from leave, I went flying with him again and all went well. After the flight he said something along the lines of "I heard you didn't get on very well last week but there was nothing wrong with that at all". Confidence restored and a valuable lesson learnt about an instructor's ability to instil or destroy confidence in a student.

One day I had lunch in the aircrew 'greasy spoon' and talked with Charlie Ashe who had been on one of the graduate entries at Cranwell at the same time as all of us from 100 Entry. A little later I was in the students' crew room when the station tannoy announced a 'State 1 – Crash'. We all rushed outside and could see a pall of smoke from beyond the far side of the airfield. Charlie Ashe and another solo student had flown into each other in the circuit in two Hunters and the aircraft had landed in a caravan park causing further casualties on the ground. Neither of the pilots had ejected. I was standing next to Ross Payne who said, "I've seen a few of these in my time and my first thought is always thank God it isn't me", a sentiment we would all experience many times in the years ahead. Following the crash and with flying cancelled most people headed back to the officers' mess and then off to the beach. As duty student I had to stay behind to tidy the crew room and clean the coffee bar before walking back to the mess. When I arrived there was a telephone call for anyone from my course; it was one of the instructors saying that we were resuming flying and I had to get anyone still around back to the squadron. In the event, there were only two of us and we were sent off on solos with instructions to ignore the crash site. Just getting on with it – the stiff upper lip – was the military way of coping. I remember walking out to the aircraft, remembering lunch with Charlie who was now dead in the wreckage of a Hunter on the far side of the airfield, but once I got into the aircraft my mind was so occupied with the task in hand that such thoughts were quickly pushed to one side.

Carrying out a dual high-level formation flight later in the course, I was in a formation of three aircraft in close attendance when I saw something pass just underneath us. My instant thought was that it was a bird but you don't get many of those at 30,000 feet! At the same time my instructor said, "Sh*t that was a formation of three Hunters that just went right underneath us!" That event nearly took out six aircraft and 12 pilots. Needless to say, procedures were tightened up immediately after this occurrence to improve separation between formations and to avoid a repetition. An exciting, but this time planned, event on the course was the opportunity to go supersonic. Whilst the Pilots Notes for the Gnat said that this could be achieved in a shallow dive, in reality this entailed climbing to 45,000 feet over the Irish Sea, rolling inverted, pulling quickly down to 30-degree pitch, rolling the right way up again and maintaining a full 30-degree nose down to persuade the aircraft to pass Mach 1. From inside the aircraft there was little to note apart from the hesitation on the machmeter as it stopped at about Mach 0.98 before jumping to Mach 1.02.

On my final navigation test with Squadron Leader Al Holyoake we flew a high-low-high profile into southern Scotland. Having found the low-level entry point OK, I set off on the route which comprised a number of insignificant points such as minor road junctions as the turning points. I got to the end of the first leg, couldn't see the turning point but turned on time (as we were expected to do) and continued. This happened with every turning point! I felt pretty stupid when we got back for the debrief when Al said, "you didn't see any of the turning points did you?" to which I had to reply in the negative. He then said I had been on track the whole way round and couldn't fault my navigation and gave me a very respectable mark for the test.

My last flight on the course was my night-flying test with Roy Gamblin. In the pre-flight brief he told me I was to start by climbing as high as possible (ignoring the 45,000-feet limit in the Pilots Notes) and carry out a supersonic run. Thinking he might be trying to catch me out, I said we couldn't do that as it required an aerobatic manoeuvre (rolling to the inverted and pulling down) which was not permitted at night due to the risk of disorientation. He told me to ignore this as well and this is what we did. We had a great ride culminating, inevitably, in a simulated hydraulic failure and recovery back to Valley in manual flying control which was one of the most demanding approaches on the Gnat (especially at night).

As we approached the end of the course, thoughts became finely focused on our postings. Whilst we had all put in our bids, our actual postings would be largely driven by our instructor's view of our abilities and the training slots available on front-line aircraft. Whatever posting we got now would determine at least the next three years of our lives and, in many cases, the direction of a whole career in the RAF. At this time (1972), the top aircraft to go to was the Harrier and the first first-tour pilots were now being trained on this revolutionary and exciting aircraft. In the event, the two top pilots on my course, Paul Hopkins and Bob Mason, were posted to the Harrier and, as third on the course, I was posted to the Central Flying School (CFS) to train as a qualified flying instructor (QFI). Having shown an interest in this posting, not least because I was invariably impressed with the instructors I had flown with in training, I was very happy; not least, as long as you did a good job, first-tour QFIs (known as 'creamies') normally got their choice of posting at the end of their instructional tour. Before leaving Valley, I was summoned to the chief instructor's office, not normally a good sign. In the event it was good news: I had been awarded a 'pass with credit' on completion of my RAF training (as had Paul).

CHAPTER 3
FLYING INSTRUCTOR 1972–75

In September 1972 I arrived at RAF Little Rissington in Gloucestershire to commence the six-month instructors' course. There were four of us creamies on the course but we were heavily outnumbered by the experienced pilots moving to instructing from their front-line aircraft types. I was allocated to the Jet Provost; my fellow trainee instructors came from Canberras, Lightnings, Nimrods, Vulcans, a pilot returning from loan service in Oman and a pilot who had come to train with us from the Singaporean Defence Force. So it was back to ground school again, not only to re-visit all the basic knowledge but also to learn how to impart it to our future students. Once we got to the flying, it was back to the Jet Provost Mk. 3 which, after the Gnat, was something of an anti-climax. However, after the initial quick conversion back onto the aircraft, as the emphasis turned to teaching, the course became more interesting and demanding.

Throughout its history, the RAF had been using the method of flight instruction developed by Major Robert Smith-Barry in 1916. This had become a set of flight 'sequences' which required the QFI to deliver them virtually verbatim. As we started our training at CFS we were told that rather than being given the set of sequences to learn, we would be given the required outcome of each teaching exercise and we would have to devise our own sequences to achieve the learning objectives. This was revolutionary for the Central Flying School but I had steady guidance and a calm helping hand from my instructor Arthur Cattle, an ex-Vulcan pilot. In between flights, I was introduced to the intricacies of 'Uckers' and 'Acey Ducey', two excellent crew room board games that aircrew used to while away the hours between flights. Our games of Uckers often lasted so long that you could handover your place to another course member whilst you went flying and return to the game when you got back to the crew room two hours later.

Early on in the course I had an unexpected introduction to monitoring a 'student' in flight when I was carrying out a mutual general handling flight – when two student instructors fly together to practise what they have been taught – in a Jet Provost Mk. 3 with one of my fellow student instructors. John Mazzurk was a transport pilot by background and had just returned from flying Skyvans and Caribous on loan to the Sultan of Oman's Air Force; consequently, he had been away from light aircraft flying for some years. John was carrying out a glide approach and a roller landing (i.e. a touch and go) on the very short runway at Little Rissington. The approach was not well judged and it was obvious to me that we were going to land well down the runway which would not leave enough room after touchdown to either take off again or to stop. Being a very inexperienced pilot flying with a much more seasoned operator, I did not feel the need to point this out to John. However, as he continued the approach, I realised that he was set on carrying out his plan of landing then taking off again. By the time we got to the flare before landing, I felt I had no option but to take over control and select full power as John belatedly realised how little runway was left ahead of us. One of the negative characteristics of jet engines in such a situation is that, unlike piston engines, their spool-up time is protracted. Consequently we cleared the end of the runway and the Cotswold drystone wall off the end of it with very little clearance as the engine slowly wound back up and we staggered into the air.

The course progressed satisfactorily although, looking back, I fear I was probably a somewhat over-confident and immature trainee flying instructor. Nevertheless, I graduated as a B2 (i.e. probationary) QFI in March 1973 and moved to No. 1 Flying Training School at RAF Linton-on-Ouse in Yorkshire to start my first tour. Although I was still a pilot officer on the CFS graduation photograph I was promoted to flying officer immediately before arriving at Linton-on-Ouse, otherwise I would have had the distinction of being a productive pilot officer QFI which by then was unheard of in the RAF. After passing through further scrutiny in Standards Squadron, I was assigned to the staff of 2 Squadron which was an eclectic mix of experienced pilots re-roled from the front line into instructing duties. They were a wonderful group of characters who made me feel welcome and helped me develop as an instructor, as a military pilot and as an individual. My flight commander was Ian 'Luigi' Chalmers who had flown the early jet fighters and had crashed two Meteors in the space of two weeks on his first tour. Fellow QFIs I remember well from my time on 2 Squadron include Bob Weetman, Mike Sykes, Bill Lewis, John Anders, Mike Kempster, Angus Morris, Neil Buckland, Duck Webb and Barry Holding. One of my first students was a university graduate and older than me by a number of years (I was still only 21 years old); he was also more senior in rank, being a flight lieutenant due to his university degree, whilst I had only just been promoted to flying officer. Nevertheless, as I was an instructor, in the disciplined flying training environment he was required to address me as 'sir' at all times.

I enjoyed instructing and had a happy and successful two-and-a-half years at Linton-on-Ouse. After six months I upgraded to B1 (average) QFI although my upgrade back in Standards Squadron did involve one unfortunate incident. I was flying with a venerable and very experienced Standards QFI who had not accepted that Central Flying School was now giving its students the autonomy to devise their own instructional sequences. Consequently, when I started trying to instruct him in the air, he flew into a major rage and started shouting at me that he was not happy with the content of my airborne lesson (which had been approved by Central Flying School). I turned to look at him and still remember his H-type oxygen mask bouncing off his face as he berated me in no uncertain terms. Despite his huge experience as a pilot and as an instructor, and having little humility at this stage in life, I decided I was not prepared to take this from him. Not least, neither at this time nor at any stage in my future flying career, did I ever think that shouting at someone in the air would ever help to produce the required results. Since I was flying the aircraft at the time, without a word I just turned the aircraft back towards base, rejoined the circuit and landed without ever speaking to him (even though he was the aircraft captain). As we taxied in, I replaced the pins in my ejection seat (not permitted until the aircraft had been shut down), taxied to our parking spot, shut the aircraft down and walked off to Standards Squadron leaving the instructor to sign the aircraft back in at the line hut. Once in Standards I went to see the officer commanding to explain what had happened and to tell him that I had no wish to fly with that instructor again. This was agreed once the other instructor had given his version of events. I flew with OC Standards the next day, completed my B1 work-up with other instructors over the next two days and passed my B1 upgrade with the chief instructor the following week. The run-in with Mr Grumpy in Standards was, to a large extent, offset by flying with a lovely instructor called Mike O'Dwyer an ex-Canberra pilot. Shortly afterwards he left Linton-on-Ouse to undertake the test pilot's course at Boscombe Down. I was very saddened to hear he had been killed whilst on the course flying an asymmetric approach in a Meteor.

I had mixed success with my students; the 'chop' rate at that time was quite high and undoubtedly somewhat higher than my course at Cranwell. One of my more successful students went on to fly helicopters and enjoyed a good career on search and rescue. Another wanted to go on to fighters and we worked hard together to get him to RAF Valley which we just managed. To an extent unsurprisingly, since he was a borderline case for the fighter stream, he failed the Gnat course about halfway through. Unusually, and most unfortunately, at that time there was no option to revert to 'heavies' where he would have done well (the 'heavy' stream being oversubscribed). If he had not done quite so well at Linton, he would have gone direct to multi-engine training and would have become a productive pilot. In the event he re-roled to navigator and enjoyed a successful career on Canberras. But what a waste of all that pilot training – and a waste of a trained pilot who was actually quite competent. I also recall one of my earliest students who showed little enthusiasm for flying which manifested itself in a lack of pre-flight preparation (e.g. learning the limiting speeds for the flaps and undercarriage etc.) and no interest in performing aerobatics. I was so surprised by his lack of commitment that I spoke to the flight commander after only a few trips with this student to voice my concern that we were wasting our time trying to train him. I was told to persevere, which I did, and in time the student demonstrated some ability but it was evident that his heart was not really in flying. It was right at the end of the course, after a year's training, that he finally got the chop.

We had a number of foreign students from Africa and the Middle East train with us during my time at Linton-on-Ouse. Whilst the RAF students had all been through a rigorous selection process and had passed Elementary Flying Training on the Chipmunk, this was not always the case with some of the foreign students, which could lead to some interesting episodes. Some of these issues related to flying aptitude, others were more cultural. A year into my tour, I had a foreign student who I started instructing from the beginning of the course. It was hard work getting him through the early exercises as he showed little aptitude for flying although, eventually, I did get him to a standard where I considered it was worth progressing to flying circuits which was the precursor to sending him solo. At this point I went on leave and handed him over to my good friend Bob Weetman. Returning a week later, I asked Bob how the student had done. Bob said he had made little progress although, towards the end of the week, the student had managed to fly nearly all the way round the circuit without any intervention; he had rolled the aircraft out on final approach, albeit rather low, about 200 feet instead of 300 feet. Always keen to encourage, Bob said, "Great, now the flap," meaning select full flap. At this point the student grabbed the flap lever and selected fully up, thereby removing a whole load of lift from the wings such that the aircraft sank even lower into the undershoot with Bob desperately selecting full power and climbing away about 100 feet above the ground. That was the end of the student's flying training so I never got to fly with him again which was something of a relief.

Another occurrence involved one of the foreign students who never performed well on his morning flights but flew noticeably better in the afternoons. The student was asked if he was eating breakfast (low blood sugar level from lack of food in the morning being a likely cause). "No," he said, "I don't like your English breakfasts". His instructor asked him if he was coming into work having not eaten or drunk anything. "Oh, I always have a drink in the morning," he said. The instructor asked what he had to drink to which the student responded, "A couple of fingers of whisky". He had obviously never been told that alcohol and flying don't mix!

All students were required to undertake a one-week survival exercise in the Yorkshire Dales during the course and on a number of occasions I volunteered to go out as one of the directing staff. A week walking in the hills was right up my street and, not least, whilst the students were surviving overnight with just a parachute and minimum rations in a wood, we would be in a Dales pub enjoying a good meal and a few beers. Setting out on one of these weeks, having been dropped off in Wharfedale, the first part of the day's walk went straight up a steep hillside out of the valley. Very quickly the only foreign student in the group started dropping back and as most were reaching the top, he was only halfway up, stopping frequently and then he started being very sick. When we checked whether he had eaten that morning, he assured us he had. It turned out he had eaten five Mars bars because he knew that they gave you energy!

About halfway through my tour at Linton I was sent off on my 'creamie's benefit'. Since creamies had little experience of the wider air force, this was a two-week attachment to one of the transport squadrons to see the real air force at work and to experience an overseas flight. I was attached to a Britannia squadron and allocated as supernumerary crew for a ten-day flight out to Singapore and back. The crew were a really good bunch of guys and looked after me very well. I remember that some wag had written on an electrical junction box on the flight deck 'in the unlikely event of the co-pilot of this aircraft performing any task correctly, this box will automatically dispense a banana'. We made it out to Singapore OK despite a gruelling final leg at night through thunderstorms on the Inter Tropical Convergence Zone which we could not fly over, the Brit not being able to get above about 24,000 feet. On the return to the UK we went unserviceable at Gan in the middle of the Indian Ocean. With no prospect of the Brit coming serviceable before I was due back at work, I was allocated a seat on a VC10 that was routeing through Gan on its way back to the UK. This heralded a requirement for a farewell party with the Britannia crew on my last night with them. Naturally, this meant I got to bed in the early hours of the morning, somewhat the worse for wear, with the result that I slept through my 0500 alarm and woke to one of the RAF Movements staff banging on my door saying the VC10 was about to depart. He took me direct to the aircraft which was sat on the apron with the engines running. I ran up the steps with my bag to be met by the head of the cabin crew, a warrant officer, who greeted me with, "Welcome aboard, sir, you've chosen the right aircraft to be late for, the squadron commander's flying this one". Ignoring the glares of the other 100 passengers or so, I shrank into my seat and suffered my headache in embarrassed silence en route to Akrotiri (Cyprus) and then back to Brize Norton.

A break from the routine of instructing came in June 1974 when we were tasked with flying a 16-ship of Jet Provosts for 'Skywatch', a one-off live BBC programme that showcased the Royal Air Force. For this we detached to Cranwell for three days with rehearsals each evening and flew back after the actual broadcast on the last evening. I flew with my very good friend, Bob Weetman (ex-Shackletons), one of the nicest and most amusing people with whom I have had the pleasure of working and socialising. Very sadly he died of a heart attack some years later in his early 50s. Whilst returning to Cranwell was, for me, almost like going home, a number of my fellow instructors had never been there (having completed officer and basic flying training through

Skywatch pre-launch, June 1974. Author centre with Ted Gould (seated). Bob Weetman is standing behind Ted.

another route) and they saw it as something of a mythical place. They were keen to be shown around and I was happy to oblige. Their reaction to College Hall and the excellent facilities reminded me what a privilege it had been for us to complete our training in such a wonderful environment.

In October I passed the upgrade to A2 (above average) under examination back at Little Rissington with the officer commanding Examining Wing. Shortly after this upgrade, the chief flying instructor at Linton-on-Ouse advised me that my A2 upgrade report had suggested that I could be considered for an upgrade to A1 (exceptional) if I wished to do this; alternatively, I could move to Standards Squadron for the last six months of my tour to assist with the checking of other QFIs. Having just spent months of personal study time preparing for my A2 I was in no rush to spend my final six months at Linton doing more of the same to gain an A1. In addition, my own view was that creamies could not realistically be assessed as military A1 QFIs since they had no front-line flying experience. Consequently I elected to move into Standards Squadron and had a very interesting time checking the flying standards of other flying instructors, the majority of whom were much more experienced as pilots but less experienced as instructors than I was. I also still flew with students on occasions, mostly on flight tests.

One of the members of Standards Squadron was Flight Lieutenant Stan Witchall, the station navigation officer, who always spoke very fast, and one day he told me the story behind his rapid diction. He had been a wartime Spitfire pilot in the desert and

Skywatch returning to Linton-on-Ouse.

he was taking an aircraft from one base to another as a singleton. The engine failed and, being unable to get it restarted, he thought he would parachute rather than risk a forced landing in the desert dunes. Having unstrapped and started climbing out, he then decided he did not fancy trusting his life to a parachute so got back in the cockpit to carry out a forced landing. He did not have time to strap back in and in the subsequent crash landing, he was knocked unconscious. Waking up a short while later, he could smell fuel and was very aware of the high risk of the aircraft going up in flames, so he leapt out of the cockpit and ran away across the sand. Once well clear he threw himself onto the desert and waited. Nothing happened. He then realised that on leaping out of the aircraft, he had broken his ankle and couldn't walk – it was only the adrenalin that had overcome the pain and enabled him to get away from the aircraft. After a while, he crawled back to his Spitfire to get the survival kit including the water which was obviously essential; this was stored in the wings but when he opened the wing panels, there was nothing there. Although he would soon be reported missing, finding him and his aircraft in the middle of the desert was unlikely and he reckoned his only realistic chance of survival would be to make his way north to the nearest friendly forces. When darkness came he started crawling; he crawled all through the night. At dawn he looked back and could still see his aircraft, still very close. Very soon after this he became incapacitated. He awoke with some Bedouin tribesmen pouring water on his lips and they took him to the nearest British Army troops. He was medically examined by an army doctor who said that apart from dehydration, a bump on the forehead from the crash landing and a broken ankle he was in remarkably good shape, but had Stan always spoken that fast? It appeared that his speech had been affected by the traumatic experience and this was still with him some 30 years later.

As a relatively inexperienced pilot, one of the things I particularly enjoyed about my tour as a flying instructor was the responsibility you were given and the calls on your integrity; this was even more so once you became a Standards instructor. I particularly recall doing a check ride on an elderly and experienced squadron leader (at the time I was a newly promoted flight lieutenant). This should have been a formality but unfortunately it wasn't and I had to take the decision to fail the check and to pass the matter on to OC Standards Squadron to resolve. We also had a new QFI posted in who had barely met the grade at CFS and was passed on to Linton with a warning about his limited abilities. After some scrutiny he was passed through Standards but a number of years later he most unfortunately came to attention having had a flying accident as a front-line squadron commander (then serving as a wing commander), killing himself and his crew. I spoke to a navigator who had known this officer when he had been a flight commander (squadron leader) and he said that his reputation as a pilot had not been good. This was the first time (but not the last) when I would come across a case of an officer progressing through the ranks on his abilities as an officer despite his shortcomings as an aviator. Given the pressure on senior aircrew officers in flying command appointments, this was not a good combination, although I hasten to add that, in my experience, by far the majority of senior officers in flying appointments had very good flying skills.

CHAPTER 4

TACTICAL WEAPONS UNIT AND HARRIER CONVERSION 1975–76

After my two-and-a-half years at Linton I was keen and ready to move on to the front line and was delighted to get my first choice of posting to the Harrier. After fast-jet refresher flying at RAF Valley on the Hunter, I moved to the Tactical Weapons Unit at RAF Brawdy in Pembrokeshire for a six-month course to learn the rudiments of fighter flying – tactical formation, air combat, air-to-ground weapons delivery, air-to-air firing and simulated attack profiles (SAPs). It was wonderful to fly the Hunter, an outstanding fighter that had been the backbone of RAF Fighter Command through the late 1950s and 1960s. By this time the aircraft were certainly beginning to show their age and I recall landing off my first Hunter solo at Brawdy with three unserviceabilities on the aircraft.

I also remember leading my flight commander, Pete Griffiths, on a pair's navigation flight into Devon in marginal weather conditions with low cloud, rain showers and poor visibility. After some time, I became 'temporarily uncertain of my position' (aka lost – you can get lost very quickly when you're travelling at seven miles per minute) which, for an ex-creamie was something of an embarrassment. After a few minutes I crested over a ridge line in the gap between the hill and the low cloud, with Griff hanging on to the back of my aircraft, and suddenly I knew exactly where I was as there was Exeter Airport right in front of us! A quick turnabout to get out of there and I managed to complete the rest of the flight without any more embarrassment. Back on the ground I taxied back in feeling very foolish, went off to make the coffees and braced myself for the debrief with Griff. I opened by saying, "OK, I have to admit that I had no idea where we were until I saw Exeter right in front of us," to which Griff replied, "Don't worry, mate, nor did I." – what a wonderfully honest guy. Battling against poor aircraft availability and the abysmal weather factor in south-west Wales over the winter of 1975–76, my course left Brawdy without ever completing all the flying events. However, since I still had some weeks to go before commencing my Harrier training, I was sent to join 45/58 Squadron, the last-ever front-line Hunter squadron which was based at RAF Wittering, to consolidate my fighter lead-in training.

Flying with some of the 'old and bold' Hunter pilots on 45/58 was something of an eye-opener. Going into some very heavy rain at low level (250 feet) in Gloucestershire one day in a two-seater (Hunter T7), the forward visibility dropped to zero. The flight commander I was with was flying the aircraft and calmly explained that since he knew the area well and that there was only one vertical obstruction anywhere near us (and that was off our route), we could keep going by looking out of the side and maintaining our height (no radalt on a Hunter) until we came out of the shower. On another occasion, I was flying a single-seat FGA9 on a low-level navigation exercise with the squadron commander following me in another single seater. Encountering very bad weather in the Welsh borders I turned around to get away from the higher ground and to consider our options. At this point the squadron commander called that he was overtaking me and for me to follow him. We then bored into the worst of the weather in very poor visibility with condensation trails forming on the vortices off the wings in the rain as we twisted our way through the valleys and eventually into better weather where he handed me the lead again. Exciting stuff!

There was a rather amusing incident during this time involving 'Smokey' Green, a USAF exchange pilot who was also flying with 45/58 Squadron whilst waiting to start on the same Harrier course as me. Having arrived from the US, Smokey had completed the UK familiarisation course on the Hunter at RAF Brawdy but was still not that familiar with the topography of the UK or the low-level environment. But one fine afternoon it was decided that Smokey would be given a Hunter FGA9 (which carried lots of fuel) and he was dispatched on a high-low-high navigation profile into Scotland. When the time came for Smokey to arrive back at Wittering there was no sign of him and I was at the squadron operations desk as air traffic were asked if he was on frequency – he wasn't. The area radar unit, Midland Radar, was then asked if they had him – they hadn't. The airfield closing

Author with Hunter FGA9 of 45/58 Squadron at RAF Wittering in April 1976.

time was fast approaching and a request was made to keep the airfield open. The squadron commander and one of the flight commanders took over the operations desk and the rest of us were told to make ourselves scarce. Eventually, Smokey did pitch up on Midland Radar and, although doubts were expressed about how much fuel he had left, he did make it back to base. The debrief didn't take long. Asked where he had been, Smokey said that he found his low-level entry point OK but after that he had flown around Scotland for the next hour or so but had no idea where he had been. Even to this day we don't know whether the Queen had an impromptu flyby at Balmoral that day. After a couple of weeks of this excitement it was time to start preparing for the mighty Harrier.

Conversion to a new aircraft type always involved a two-day visit to the RAF Aviation Medicine Training Centre at RAF North Luffenham to undergo refresher aviation medical training, be issued with the aircraft specific aircrew equipment and undergo an explosive decompression (normally 25,000 ft to 45,000 ft in three seconds) to simulate the loss of pressurisation and to experience the effects of hypoxia (lack of oxygen). As aspiring Harrier pilots, we also spent a week at RAF Ternhill undergoing helicopter training on the Whirlwind to practise hovering and the transition to and from the hover. Reporting to the helicopter training squadron on Monday morning we had no idea what was in store; we imagined that we would spend the morning being briefed on procedures and techniques and might get to fly that afternoon. Around 0900, one of the QHIs (qualified helicopter instructors) came in and asked, "Which of you is Chris Burwell?" I jumped up and was led out to a Whirlwind, told to climb up and get strapped in. The QHI then did an external check of the aircraft, got in the other seat, started the engine, did some checks then asked air traffic for permission to lift off, all without saying anything to me. Once we were cleared, he then hover-taxied the aircraft to a clear area and then told me that I had control – and that set the scene for the week. We were not required to know anything about how the aircraft worked, air traffic procedures or radio calls, all we had to do was fly the aircraft which was immense fun. The QHIs enjoyed it too as, by the end of the week, they would have us doing very advanced exercises which tested us to the limit (quick stops, turns around the tail rotor on the airfield, landing into clearings and turns around the tail rotor in the clearings!). Twice that week I was pushed beyond my capabilities by my QHI and had to hand back to him in a hurry before it all went horribly wrong. After that it was time to return to Wittering to start two weeks of ground school and simulator and then to actually fly the Harrier.

Over the preceding few months, four of the instructors on the Harrier Operational Conversion Unit had been killed – one had flown into a mountainside in Norway, two in a mid-air collision and one when he failed to put his ejection seat pins back in the seat at the end of a flying display and inadvertently fired the ejection seat on leaving

No. 19 Harrier Course 233 OCU at RAF Wittering, May–November 1976.

the cockpit. The Harrier had been involved in a number of accidents during its limited time in service and quite a number had been fatalities. Before starting the flying, Ted Ball and I, the only members of the course living in the officers' mess, had a discussion along the lines of 'learning to fly this aircraft is going to be very challenging, we are going to treat our training and the aircraft with huge respect and, to this end (and to improve our chances of staying alive to the end of the course), we will not drink any alcohol during the week'. And we didn't; though we tended to make up for it on Friday nights.

Learning to fly the Harrier was a tremendous experience – exciting, very challenging, at times it could be quite frightening but ultimately hugely rewarding. The initial conversion comprised 21 flights which covered all forms of take-off and landing and, towards the end, included vertical landings onto a concrete landing pad in a wood off the end of the airfield, vertical landings onto a small MEXE (metal planking) pad, and landings onto a grass strip at another airfield – these latter flights all solo without the benefit of a dual beforehand. The learning curve was incredibly steep and we were only permitted to carry out a maximum of two flights per day, even if these were very short vertical/hovering exercises which were logged as 15 minutes but would normally last well under ten minutes airborne. We had just two duals before going solo followed by a hovering dual then a hovering solo; so your fifth flight in the aircraft was three solo hovers (or press ups). The summer of 1976 was very hot and we would often finish flying by early-mid-afternoon and would go to the tennis courts to unwind. However the high temperatures also had an impact on the hover performance of the Harrier which gave rise to one interesting incident during our training. One morning we were all due to carry out our MEXE pad landings. There being no dual for this, the psychological pressure of convincing yourself that you could safely decelerate to the hover and put the aircraft down on a very small landing pad that you could not see once you were in the hover was huge. That morning there was only one aircraft available for our course for this exercise so we all had to wait our turn. Starting alphabetically, Ted Ball went out first and the rest of us watched in some trepidation from the crew room, as our turn would be coming up soon, to see how Ted got on. On his first approach to the pad all was looking good until shortly before he got to the hover when he suddenly accelerated away, went round the circuit and came in for another go. This was a bit of a surprise as the weight calculations had shown that, despite the high temperature which had the effect of reducing hover performance, the aircraft should have been well within its hovering performance. Nevertheless, Ted got down OK on the second attempt and completed the sortie satisfactorily. Being second to go, I checked with Ted what had happened on the first approach and he told me that the '15 second' light had come on as he approached the hover; this indicated that he was running out of power. On my first approach to the pad, with the adrenalin level sky high (not helped by the fact that I knew the rest of the course would be watching me), the '15 second' light came on for me as well which, at this stage of training, was quite unnerving. Today was the first time any of us had experienced this situation and an overshoot at a late stage in the deceleration was something we had not practised. Anyway, all went well and, like Ted, I managed to get down on the pad at the second attempt. Next out to fly was Gavin Mackay who managed to do something which neither Ted nor myself had done – he read the aircraft technical log (Form 700) carefully; this stated quite clearly that the aircraft was carrying 50 gallons of water (for turbine disk cooling) and was therefore 500 lbs heavier than we thought. This explained why the aircraft was reluctant to go into the hover. Once Ted and I had overshot and flown round the circuit we had used enough fuel to bring the weight of the aircraft down to hover performance. The water should not have been in the aircraft but Ted and I should have picked this up from the F700 so we were both ultimately responsible for putting ourselves in a somewhat precarious predicament. A valuable lesson learnt the hard way.

On the completion of the first 21 flights, as with every other course, we were shown 'The Horror Film'. At the start of the Harrier's life in RAF service, there was no two-seat training aircraft so all the early training flights before the advent of the two-seater T2 were filmed. This provided a very useful means of debriefing pilots on their early attempts at hovering etc. but now provided a record of what could go wrong if the aircraft was mishandled. Having watched some entertaining attempts at first vertical take-offs and loss of directional control on the runway during a conventional landing amongst others, the film finished with a loss of control in an accelerating transition from the hover. The pilot ejected, too late, and was killed on impact with the ground. It was a subdued group of seven trainee pilots that left the room. There was a good reason for not showing the film until the initial conversion had been completed: we now knew how to do every type of take-off and landing in the aircraft but this was an incredibly powerful reminder that this aircraft could kill you if you did not respect it. If we had been shown the film earlier on, we would probably have been even more concerned for our own safety than we were during the conversion. Having converted to the Harrier with the luxury of a simulator, the T4 and the GR3 (both with uprated Pegasus engines from the T2 and GR1), I can only express my admiration for Dick le Brocq, Bruce Latton, Peter Dodworth and Richie Profit of the original Harrier Conversion Team and all those who converted to the aircraft prior to the introduction of the T2.

After the initial aircraft conversion, we progressed onto navigation exercises using the inertial navigation system (which was a relic rescued from the doomed TSR2 project), instrument flying, air combat and night-flying qualification before moving on to the advanced phase of training learning weapons delivery (bombs, rockets and guns), reconnaissance techniques and attack profiles. The conversion course culminated in Exercise Tartan, the final week of training being undertaken in northern Scotland with attack profiles in the Highlands and first run attacks dropping practice weapons on Tain Range near Inverness. On my last flight I had the boss of the OCU chasing me in another single-seat aircraft. Approaching the first target – a line search along a road – the weather deteriorated and halfway along the line I decided the weather was unfit to continue and pulled up above the low cloud into clear air. The boss quickly voiced his view on the radio in no uncertain terms that the weather had been acceptable and I should not have pulled out. Continuing above cloud on heading and on time, some minutes later I was able to descend back to low level but now I felt I had a point to prove. Consequently, when the weather deteriorated again, I continued at very low level with the cloud just above us (well below the stated minimum criteria for low flying) and pressed on to Tain to carry out the planned strafe attack[5]. This seemed to satisfy the boss and I had completed my Harrier conversion, being rewarded with the Ferranti Weapons Aiming Trophy and a posting to No. 1 (Fighter) Squadron based at RAF Wittering.

CHAPTER 5

NO. 1 (FIGHTER) SQUADRON: THE NORTHERN FLANK OF NATO AND THE REINFORCEMENT OF BELIZE 1976–78

I commenced my first front-line tour in November 1976 joining the squadron with Chris Gowers, a first tourist, who had been on the OCU course with me. Our first few months were taken up with reaching combat ready status which, for No. 1(F) Squadron with its responsibility for supporting the northern flank of NATO (Denmark and Norway), included learning air-to-air refuelling. During this early period the squadron deployed to Tromsø in northern Norway in winter, a field exercise at RAF Leeming, a squadron exchange with the Belgian air force and a NATO TACEVAL (tactical evaluation) in Denmark. It was a busy time and, once again, a steep learning curve for us new Harrier pilots. I can still recall hanging on to my leader as we carried out a hard turn through 180 degrees in a valley in Norway on the edge of a snow shower; as we turned through 90 degrees all I could see was a huge mountain disappearing up into the low cloud very close on the nose and me thinking 'I hope he's got it right and we have enough room to turn round in this valley'. On the field exercise I remember getting airborne after a 'cockpit turnround', which involved replanning in the cockpit (the first time I had done this), and picking up my map once airborne to find I had omitted to put any headings on it which provided Bernie Scott in the back seat of the T4 we were in with much amusement. The less said about the social pressures of the squadron exchange in Belgium the better.

The TACEVAL in Denmark in June would have been a wonderful camping holiday – the weather was sunny and very warm, the company and the food was great, the tents and beds were almost comfortable, the only problem was that we were being assessed by a team of our peers from across the NATO nations. The flight profile that we were given by the TACEVAL team proved to be very, very tight on fuel (a commodity that was always pretty short in the Harrier GR3/T4). Given the good weather, and the availability of a large civilian airport as a diversion on our flight path as we returned from our missions, a decision was taken that the squadron would not raise an objection with the TACEVAL team and we would press on as tasked (typical

No. 1 (F) Squadron Harrier GR3s deployed to Tromsø, Norway.

Harrier Force!). On one of my missions, I was given the T4 to fly with a TACEVAL assessor in the back seat (a USAF major). He seemed very pleasant and we had a good chat as the flight progressed but as we pulled up to high level in the north of Denmark to get back to base (to use the minimum amount of fuel), he went very quiet and I hardly got a word out him all the way home. After landing he went off without saying anything, leaving me wondering what I had done wrong. That evening an RAF squadron leader, a Harrier pilot who was working with the TACEVAL team, looked me up and asked how I had got on with the USAF major. I told him everything had been fine until the last 20 minutes, to which he said that when we pulled up to high level in the north of Denmark, the major was convinced we didn't have enough fuel to get back to base and he thought that we were going to have to eject.

Somewhat to our surprise, our camping holiday was cut short due to developments in Central America. The Guatemalans were agitating once again to gain access to the Gulf of Mexico through a land claim on Belize (formerly British Honduras and, at the time, a British protectorate). We returned to Wittering and began preparations for a deployment to Belize. Since Belize had just the one 6,000-ft runway and only jungle strips for diversion airfields, and the Royal Navy's fixed-wing carrier-borne force was now out of service awaiting the introduction of the Sea Harrier, the RAF Harrier was the only fighter aircraft that could be deployed to meet this threat. As the primary role would be air defence against any incursions by the Guatemalan air force A38s, we now had to focus our attention on air-to-air firing using the Harrier's Aden cannon. Assisted by one of the Phantom Force's qualified weapons instructors from Coningsby, we were sent off to shoot against 'the flag' – a target towed by a Canberra aircraft. Although the Harrier had now been in service for eight years, the air-to-air gunsight had never been calibrated and no-one knew where the bullets were going relative to the aiming pipper. Consequently, when we went to fly, we were given different aiming points to try and work out a solution that would put a bullet on the flag. This was eventually achieved on the Wednesday of that week. For those not familiar with air-to-air firing, it is probably worth pointing out that this was all incredibly difficult. Even with a gunsight that is giving the correct computation, meeting the correct firing parameters in terms of speed, range and angle-off the flag (critical for safety reasons for the flag-towing aircraft) is only the start; the pilot then needs to ensure that the pipper is in exactly the right place and his aircraft is steady when he fires to have any chance of hitting the target. Once we had an aiming point, we were then able to put a few more bullets onto the flag although the air-to-air sight was still of pretty limited use to us. After some delay and uncertainty, the government decided to reinforce Belize to deter the Guatemalan's threat of invasion. I was nominated to fly one of the six Harriers in the reinforcement plan.

THE REINFORCEMENT OF BELIZE 1977 PART 1: POND JUMPING – 6 JULY

This account was originally published in *Harrier Boys Volume One* edited by Bob Marston.

The whole squadron has been called back early from the long weekend break. It is May in UK, a fine early summer and there is a growing problem in Belize. Although we do not know this at the time, in a few weeks' time we will fly six of our Harriers to Central America. We will do this to provide a show of military strength and intent which, it is hoped, will help defuse a growing confrontation with Guatemala.

The Guatemalans have a land claim on Belize, a British Protectorate, and it is not the first time this situation has arisen. Belize is a long way away from the middle of England and we will need help to get there. This will come in the form of Victor air-to-air refuelling aircraft and a night stop at the Royal Air Force detachment at Goose Bay in Newfoundland. Our Harriers will fly in three pairs, each pair with their own Victor tanker aircraft. I have been nominated to fly the sixth Harrier. I have been flying the Harrier for just one year; my leader, Chris, is an ex-Hunter pilot, an experienced Harrier pilot and a weapons instructor and I hold his background in high regard. Our aircraft will be equipped with huge external fuel tanks. This will ensure that if we have trouble taking on fuel across the Atlantic, or down the eastern seaboard of Canada and the United States and into the Caribbean, we will always have sufficient fuel to reach a diversion airfield.

Unusually, we will also take food and water with us for these two trips. Our normal sortie length is about one hour, on these sorties we can expect to be airborne for five to six hours or more.

The cockpit of this aircraft is ridiculously small. I am wearing an immersion suit. This is a bulky piece of clothing but it will help me to survive if I should finish up swimming in the Atlantic Ocean. On top of my immersion suit I wear a parachute harness and life jacket combined; under my immersion suit I am wearing a number of layers of warm clothing. It is not easy to move in this ensemble and reduces even further the normal limited space in the cockpit. I also have to take maps, navigational books, my officers' hat and food and water into the cockpit with me. These can only be pushed down the sides of the ejection seat. I have no room to move – and I will be sat here, strapped firmly into this Mk. 9 Martin-Baker ejection seat for the next six hours. Carefully, I carry out the pre-start checks. Mid-Atlantic would not be the place to discover I have over-looked a problem with the aircraft. Content, I signal to the ground crew that my ejection seat is now live and I am about to start the engine. I select start, wait the regulation ten seconds, then select high pressure cock on to put fuel through to the engine. Immediately the engine comes to life with a satisfying light rumble and a steady rise in jet pipe temperature. Within half a minute, the engine stabilises at idle RPM and I can check out all the systems: fuel, hydraulics, electrics, inertial navigation, TACAN navigation, oxygen, anti-G, radios, controls, flaps. Even at idle, there is a tremendous feeling of power from the engine as it shakes the whole airframe. All is well. The ground crew remove the chocks and I confirm with Chris on the radio that I am ready to go. We transfer to an air traffic radio frequency for our clearance to taxi.

We will take off on the easterly runway and we have a long way to go to the take-off point. There is so much residual thrust from the Harrier's Pegasus engine at idle that I must deflect the nozzles downwards and use the wheel brakes to keep the speed under control. We receive waves from a crowd of interested spectators in the control tower who have come to see us off on our journey. It is reassuring, and it is welcome, but most of them will have no idea what it feels like to be in our place. This is where a schoolboy dream of flying planes, and eight years of training and flying, has finally brought me. We receive our take-off clearance, Chris lines up and powers off down the runway right on time. Final checks complete, I select full power and check that all is well with the engine. With this amount of fuel on board the acceleration is noticeably reduced but, with 9,000 feet of runway ahead and using the nozzles to deflect the thrust to get airborne before we have true flying speed, it is not a concern. Safely airborne, with the undercarriage and flap retracted, I still have Chris in sight and set off to catch up with him on our appointment with our Victor tanker in the North Sea.

Less than two years ago I was a flying instructor teaching the rudiments of flight in the skies of northern England. It was a satisfying and often rewarding job but this is what I joined the air force to do: to be involved in the big issue of the day, to make a difference. Don't get me wrong, I have no wish to drop real bombs on real targets, or use the 30-mm Aden cannon in anger, nor do I wish to be shot at. But if I have to do these things, then I will. Today there is a small Foreign Office crisis and I am a very small part of the military response that will help resolve the situation. We have weapons and we will use them if ordered. I would not have got through pilot selection with the Royal Air Force ten years ago if I had had any doubts on that score.

But now I must concentrate on the task of getting my Harrier to Newfoundland. I have tucked my aircraft close to my leader's and we are sharing the sky with a surprising number of Victor tanker aircraft. Interestingly, it is very difficult to determine which is the one that will take us across the 'pond'. Once this is resolved we must take fuel straightaway, not that we need it but to ensure that our refuelling systems are functioning normally. We join the tanker on his left wing, I fly on Chris's left side. Once cleared by the tanker, Chris goes astern the right-hand wing of the Victor and I go behind the left wing. We each have our own refuelling 'basket' on the tanker aircraft. With the lights on the Victor's hose drum unit at red, I hold my position well clear of the basket and wait for the lights to go to amber. This will not happen until Chris's Harrier is coupled with the Victor and his aircraft is taking on fuel. I can look across and see Chris positioning himself close to his refuelling basket and making contact. My lights now change to amber and I close up to stabilise about six feet from the basket. Getting my air-to-air refuelling probe into the basket, and taking on fuel, is now down to my ability to fly formation.

Formation flying is one of the purest forms of flight. If a pilot has the training, the skill and the confidence, he can use the combination of flying controls and engine power to hold the aircraft in the correct place relative to the lead aircraft. Little exists in life once you are in close formation apart from the constant striving to maintain the perfect position. You do what you are told, your leader makes the decisions and you only take your eyes off him very, very briefly, to check your fuel or oxygen or to change radio frequency or altimeter setting. Pilots are often criticised for having too much confidence. If I do not have enough confidence in my ability to formate on this basket, which is shaking around in front of me as we charge across the North Sea at 270 knots, I will not be able to perform the task of refuelling my Harrier aircraft and we will not reach our destination. It also helps not to have too much imagination. If I get this wrong, I could damage the refuelling basket, or I could damage my aircraft, or I could collide with the Victor tanker and people could die. Sadly it has happened. I am confident. And I exclude the possibility of gross errors from my conscious thought processes.

When I was a flying instructor, the students were desperate to see and be taught how to fly in formation. One day, three of us instructors with our students on board, thought that we would take advantage of a short transit flight to show the students what formation flying is like. Because formation flying is potentially dangerous, military regulations state that all formation flights must be very carefully briefed before you get airborne. This is sensible. What we did was not. We didn't have a brief. The only agreement was that we would do some close formation flying in the transit and that the first aircraft airborne would be the leader. As we would be using air traffic radio frequencies in the transit, and because this was not an official, authorised formation flight, there was no question of us using the radio to control the formation. I was the second aircraft in the stream take-off and duly joined on the leader's right-hand side, the normal position for No. 2. We flew for a few minutes up to our turning point but there was no sign of No. 3 on the leader's left-hand side where I expected to see him. As we made our turn towards our landing airfield, the instructor in the lead aircraft looked towards me and appeared to wave me away. This was what I was expecting. As I turned my aircraft away from the lead aircraft, I turned to my right to ensure there were no aircraft out there. It was at this point that I discovered where the No. 3 had gone. He had decided to join in close formation on my right-hand side and was feet away from me. He was now desperately trying to get out of my way before we collided. How we missed each other I don't know. The leader had been trying to point out to me that No. 3 had joined on my right-hand side. So, whereas too little confidence in close formation will undermine the pilot's ability to perform the task, we proved (if proof were needed) that too much confidence is dangerous and could be fatal.

My refuelling probe is to my left and I have lined this up with the left-hand side of the Victor's refuelling basket six feet away. I need to line up to the left since, as I close the final few feet, the airflow around my canopy will interfere with the basket and push it away from my aircraft. With everything steady, I apply a 'handful' of power and my Harrier surges gently forward. Unseen, because I am now formating on the under surface of the Victor wing, the probe is captured by the basket and I feel a satisfying clunk as the probe locks into position in the middle of the basket. My aircraft has now taken up the not inconsiderable weight of the refuelling hose and I must apply more power to push in the hose to the orange band as this will make the fuel flow into my aircraft. As fuel flows, the amber lights go to green and my fuel gauges start to increase slowly. Now I must sit here, with the vast bulk of the Victor tanker towering over me, formating on the Victor's wing and maintaining the natural trail line of the hose. Five minutes later as my aircraft's fuel tanks reach full, I drop back from the tanker and, with a positive tug, withdraw my probe from the basket. I take up position outside the Victor's left wing and Chris sits outside the right wing.

I look down below and we are only just now crossing the east coast of Scotland. Our Victor tanker fills up his own tanks from one of the other Victors, who then turns back for his home base bidding us farewell, and we are on our way across the North Atlantic to answer the government's call. We will be heading west together at 27,000 feet for the next four hours.

There is not a lot to say about flying a single-seat, single-engined aircraft across the North Atlantic to Newfoundland apart from: it's a long way; you're very much on your own; and you start listening to the noises the aircraft makes very carefully. Three times our Victor tanker calls us in to refuel, the rest of the time I am left to my own thoughts. I listen to Jonathan Livingstone Seagull and other music on the Harrier's own on-board entertainment system (a tape recorder installed for reconnaissance missions). Although it is early July, the North Atlantic looks forbidding. It is vast and the whitecaps on the waves suggest that in the event of having to eject from the aircraft, a parachute landing in those hostile waters would be a difficult undertaking. We train every year for such an eventuality, in a pool and sometimes in the sea. The training is good, using all the right equipment; we even train blindfolded to simulate night and sometimes we go in the sea when it is very cold with our full immersion suit and kit and get lifted out by one of our search and rescue helicopters. I feel well trained to cope with the ocean should it come to that but I don't allow myself to think that it will happen. If one of us does end up down there swimming with the fishes, we have a very small one-man dinghy in our survival kit built into the ejection seat to help us survive. There is also a Nimrod maritime patrol aircraft out there somewhere, covering our route, which can come and drop us a bigger dinghy with more survival kit. But all that is pushed to the back of my mind with the excitement of the venture. Nevertheless, it is with some relief that we approach Goose Bay at last and detach from the tanker to make our individual approaches and land.

The weather is not too good at 'Goose' with a depth of cloud, an overcast at 1,000 feet, some rain and a crosswind which is a concern. The crosswind limit with the huge fuel tanks on-board is low since the aircraft handles differently with the tanks on and there is a different landing technique. I have never landed the aircraft with these tanks on. There is some confusion as Chris makes his approach as the Harrier works on true headings whereas air traffic is using magnetic headings which are very different in this part of the world. Once Chris has landed, radar give me an approach and once I am visual with the runway I am cleared to land. All goes well on the approach but on landing the aircraft has a mind of its own and heads off for the side of the runway as the crosswind does its worst on my Harrier with the big tanks. I am too slow to correct. My good fortune is that the runway at Goose Bay is very wide and, unusually, there is tarmac beyond the runway lights. It is just as well as I use this space to get the aircraft more under control and thread my way back through the runway lights, no damage done, more by luck than judgement. I park my Harrier, retrieve my kit from the cockpit and my bag from the stowage in the rear fuselage. There is a warm welcome and the chance to unwind and relax with the other pilots and the Victor crews in the bar before setting out on the second stage of the journey the next day.

PART 2: THE EASTERN SEABOARD AND THE CARIBBEAN – 7 JULY

The next day we have a briefing. All flights start with a briefing but this is different. Besides us six Harrier pilots, there are untold numbers of Victor tanker crews at the briefing as well. Each pair of Harriers will require a Victor to take them to Belize, but each of those Victors will require refuelling from another Victor at the top of climb out of Goose Bay; and then there is a requirement for spare aircraft in case of unserviceabilities. I look round and think it is reminiscent of a World War II bomber station mission brief but without the cigarette smoke and with less, but still a reasonable amount, of adrenalin – certainly on my part. At the briefing we learn that the weather is still not good at Goose Bay with rain and low cloud, a good depth of cloud to climb out through, cloud down the Eastern Seaboard and then clear across the Caribbean and into Belize. The final part of the brief is a discussion on how we take two Victors off followed closely by two Harriers so that they arrive together above cloud without losing each other. With like aircraft types we use a 'snake climb' with 30-second intervals between aircraft where the following aircraft do exactly the same as the leader 30 seconds later for No. 2 and a minute later for No. 3. With dissimilar types today (i.e. Victor and Harrier) this is complicated by the different climbing speeds and rates of climb. Eventually an agreement is reached with the Victor crews as to how this will be done.

By this time, we are aware that there is a problem with Chris's aircraft which the engineers have not been able to fix. The question now is whether I am to continue on my own or whether I will wait, perhaps for a few days, until Chris's aircraft is fit to continue. The first pair of Harriers departs with its tankers, followed by the second pair. But still we wait for a decision from HQ in the UK as to whether I go or stay. Eventually, after a long delay, the decision is made that I go.

As I walk out to the aircraft, the weather is as briefed, cool and wet. I carry out my external check of the Harrier and, much to my surprise and concern, find a large rusty nut in the engine intake, sat at the bottom of the first stage of fan blades. It is certainly not a nut that has come from that area of the aircraft. If this were to be in the intake area when I start the engine it would almost certainly cause damage to the engine that could well result in a catastrophic failure in flight. Although I report this to the junior engineering officer who is there, there is no time to delay and I never get any explanation of how the nut got there. Settled once more into the confines of my Harrier GR3 cockpit, surrounded by maps, food, water and hat, I am ready to go. I taxi out in the rain, watch my two Victors depart ahead of me then power off down the runway and into the overcast to confront the Guatemalans.

The departure is complicated, the cloud is thick and without breaks but eventually I get out of cloud and reach cruising altitude. There should be two Victor tankers somewhere not far in front of me but, despite good visibility, I can't see either of them. Back in the UK I would rely on a fighter controller to vector me onto the tanker but up here in Newfoundland, Monckton Radar don't even respond to my calls for assistance. After what seems like an age, but is probably only five minutes or so, I spot a Victor some miles ahead. I catch up with the tanker and position on his left wing; at this point, the second Victor appears from behind. So I had got ahead of him in the cloud on the departure out of Goose Bay which was not the plan at all. After some buddy-buddy refuelling between the Victors, the leading tanker turns back for Goose.

I have been hoping for some memorable views on the flight down the Eastern Seaboard but the cloud cover determines that I will not see land again until I reach Belize some six hours after leaving Goose Bay. I refuel four times on the flight south. At times we fly into thin higher cloud which means I have to fly close formation on the Victor to keep him in sight (without a radar in the Harrier, and with no radar support, there is no way for me to find the Victor again if I should lose it). Passing south of Florida, we turn onto a westerly track to head across the Caribbean for Belize, having safely navigated our way through the Bermuda Triangle. Throughout the flight, as on the previous day, I have been operating on a 'discrete' frequency with the Victor; in other words, I am not listening to air traffic. But now, as we route north and west of Cuba comes a surprise: the Victor calls me to say that air traffic will not allow them to take me all the way to Belize as planned, but they will have to drop me off some way short. They also advise me that they have no idea of the situation on the ground at the airport i.e. whether there has been any offensive action by the Guatemalans. However, if I am not happy to continue on my own, they can take me to Bermuda with them. There doesn't seem to be a lot of point in turning tail and running at the first hint of a problem, thereby denying the RAF of another Harrier in the reinforcement; not least, I wouldn't be able to face the ridicule of the other squadron pilots if everything is OK in Belize and I'm on the beach in Bermuda. Having refused the offer of more fuel (I have plenty), I take an updated position from the Victor which I insert into the inertial platform in the Harrier and set off on my own armed only with two 30-mm Aden cannon and a radio frequency for CENAMER (air traffic), who the tanker crew think may be able to help me. In the end, the drop-off is only about 150 miles out from Belize but unsurprisingly, given the abysmal radio fit in the Harrier GR3, I am unable to raise CENAMER on the radio. The next 10–15 minutes I feel very alone. But eventually, about 55 miles out from Belize International Airport, I manage to raise Butcher radar (the RAF fighter controller at the airport) on the radio who informs me that all is well on the ground. Since the aircraft is still overweight for landing, I have the opportunity to get my first look at Belize City and the north of the country before coming in for a landing that is happily less exciting than at Goose Bay.

After landing, the tower direct me to Foxy Golf Harrier Hides on the southern side of the runway: across a steel planking bridge, up a ramp and into the hides area. I shut the

> aircraft down, gather my kit and ask the ground crew if there is any chance of a lift to operations. There is no transport and they advise me to walk down to air traffic and try to get a lift from there. Picking up my bag and flying kit, I set off down the track in my immersion suit and two layers of thermal clothing, with the temperature at +30°C and in a humidity that I have never experienced in my life before.

My first night in Belize I was sharing a room with an army captain. At about 0330, someone came into our room to rouse the captain, telling him that an attack by the Guatemalans was expected at dawn and he was required immediately. I was pretty tired after two days of flying and tanking the Harrier to Belize and rolled over and went back to sleep again, knowing full well that our pilots for the morning were nominated already and someone would call me if I was required. In the event, of course, nothing happened.

The squadron held the Belize commitment for the next year at which time it was transferred to a Harrier Force shared responsibility which endured for the next 16 years; over this year, I completed four detachments of five to six weeks. Flying in Belize was most enjoyable: the country is approximately the same size as Wales with varied topography, the weather was invariably good and we had the use of an air-to-ground range and no restrictions on where we flew. (I am aware that, given

No. 1 (F) Squadron Harrier GR3s over the Blue Hole, Belize in 1977.

the difficulty of accurate navigation over the jungle down the spine of the country bordering Guatemala and the unreliability of the Harrier GR3's inertial navigation system, the odd inadvertent infringement of Guatemala's airspace did take place!) Although limited to a minimum height of 250 feet, flying lower was not going to cause a problem to anyone on the ground over much of the country which was dense jungle but you did have to watch out for the 'widow makers' – the tall trunks of dead trees that stood out above the jungle canopy which, being devoid of foliage, were very difficult to spot and had the potential to spoil your day. Our normal working day started with met brief at 0800, two waves of flying (which meant you would normally fly one trip per day), lunch, a siesta, a game of squash or a swim, afternoon tea, a game of volleyball with the helicopter guys, sun-downers (beer or a cocktail), then dinner followed by a film or a game of cards. It was a demanding life.

I recall a number of memorable flying events during my detachments to Belize. On one occasion, as I dropped the nozzles to get airborne on a standard rolling take-off, the canopy flew open with a mighty 'bang' as the mechanical latch failed to hold it closed. As I was now airborne, there was no chance of aborting the take-off nor, due to the slipstream, was it possible to close the canopy. I was now flying without the option of ejecting as the canopy arch was right above the ejection seat. Dumping fuel to get down to landing weight, I quickly put the aircraft back on the ground before anything else went wrong.

On one of my detachments, we had a visit from the Inspector of Flight Safety, Air Commodore Ken Hayr, an ex-OC 1 (Fighter) Squadron, who was current on the Harrier and wished to fly a couple of trips whilst he was in theatre. I was given the job of briefing him and leading him on two trips with a cockpit turnround in-between. The first trip was a low-level area familiarisation. On the second trip, we were going to do some practice intercepts and air combat. I briefed that after a pairs take-off in close formation, he should move into battle formation (two miles abreast) for the climb. In the event, he stayed in close formation after take-off so, realising that he wanted some practice in close, I started winding up into some big wingovers and high G turns. I have to say that his formation keeping was immaculate and at least as good if not better than most of the squadron's pilots; very impressive from a senior officer who had been away from front-line flying for a few years. It was very sad that he was killed flying a vintage aircraft on a display some years later after he left the RAF.

On another occasion, I was scheduled to fly a 1v1 air combat trip with my good friend Mike Beech. We had never flown against each other in combat but I knew that Mike was an excellent pilot so I could see that this was going to be a demanding flight. Once at 5,000 feet (the minimum height for full combat) we started the first engagement. After 20 minutes, neither of us had had a shot at the other and we agreed to stop as we were both absolutely knackered. After a few minutes break, we set up for another engagement. This lasted about the same length of time, at which point we still didn't have a result but had run out of fuel. We knocked it on the head and came back to land on minimum fuel – mutual respect.

We flew missions in support of the army, including a lot of forward air control training flights[6], and I well remember getting up very early one morning to fly on a Force exercise in the south of the country. Going into the officers' mess at 0800 we had finished work for the day. We flew in support of the SAS in theatre and practised weapon deliveries and strafing both on the range and on the 'splash target' (towed behind a ship) when the navy were around. The Royal Navy's West Indies Guard Ship didn't visit too often, normally finding much more attractive places to visit in the Caribbean rather than Belize.

One memorable period over this year was the Christmas detachment which was manned, on a voluntary basis, by the squadron's junior pilots under Pete Day the executive officer. Bob Marston, our fighter reconnaissance instructor, organised a recce competition that had some 15 targets that had to be photographed (using the Harrier's in-built port-facing F95 camera) across the whole country. We were only allowed 30 minutes planning time and, given the aircraft's limited fuel, it was necessary to use a high-low-high profile and there was certainly no option of making a second pass if you missed a target on the first run. I don't remember who won the competition but I do recall that shortly afterwards Pete Day destroyed all the photographic evidence – I think it had something to do with the height that most of the photographs were taken from. Being in Belize over the Christmas period provided the opportunity for an entertaining new year trip to El Salvador for two to three days with Chris Rayner, who was on loan from No. 3(F) Squadron and Ted Ball from IV(AC) Squadron. Using his usual charm, Chris very quickly acquired us an invitation to an ex-pat's New Year's Eve party in a beautiful house in the hills above San Salvador. Uncertain how to repay our hosts' hospitality, but knowing that fireworks were the big thing in Central America for new year, we took a load of fireworks (mostly bangers) to provide a big show at midnight. Come the hour we let loose in their lovely garden, only to discover that Salvador fireworks are mostly made of newspaper which, when ignited, shred themselves. By the time we had finished, the garden was a total mess of shredded paper. We left the party before anyone asked us to clear up the mess.

Another incident I recall was waiting to cross the runway one day in a Land Rover whilst a B727 landed. Shortly after touchdown, one of the main-wheel bogies of the 727 started smoking alarmingly as the pilot applied the brakes. Getting clearance from air traffic to drive down to the airport apron, we met the captain as he came off the aircraft. When we told him what had happened on his landing run, to hopefully avoid a repeat and possible disaster on his next landing, his laconic response was, "yeah it's being doing that all day".

Over the year in Belize, we developed a standard Friday night: a few beers in the mess then into town for a Chinese meal at the Chon San followed by a beer at the Big C, a well-known brothel over the road, before moving more into the centre of town to visit a bar or two. On a rooftop bar one night I remember listening to a local band doing a fantastic rendition of some of Santana's greatest tracks. One night we decided to spread our wings a bit and went across the Belize River and followed the music into a club. The music didn't actually stop, but it certainly felt like everyone was staring at us as we went in. It was then that we realised that we were the only whites in there. Anyway, we got a drink and everyone got on with their business. This became a regular place for us to visit on our nights in town and we seemed to be accepted there. We even took our station commander from RAF Wittering there, Group Captain David Brooke, a very urbane senior officer and a great guy – it was only sometime later that we discovered that the other side of the Belize River was out-of-bounds to all military personnel after dark (something to do with drugs and crime) – such innocence – we really should have read the force standing orders at some point.

Whilst the squadron held the Belize commitment, we used loan pilots from the Germany squadrons to relieve the pressure of the on-going detachments on our own pilots. To cover this shortfall in Germany, 1(F) Squadron was required to make good the deficit in their squadrons for their TACEVAL in the summer of 1978. I readily volunteered for this two-week field deployment and was sent on loan to IV(AC) Squadron at Gütersloh for the deployment into the Sennelager Ranges.

This was only my second field deployment (following the somewhat low-key Leeming fieldex) and, to make up the number of combat ready pilots declared to NATO for the TACEVAL, I was declared 'combat ready attack and recce' after just two sorties at Gütersloh before we deployed into the field. The day before all the pilots were assembled in the navigation planning room awaiting a field sites briefing. Squadron Leader Keith Holland (later Wing Commander Keith Holland, OC IV(AC) Squadron), who was joining the squadron for the exercise, came in to the room to prepare his sites map. All the other pilots had done this and everyone was sitting round bantering, watching Keith at work, and waiting for the briefing to begin. Having prepared his map, Keith got a couple of junior pilots to help him remove the backing off a large sheet of fablon (sticky-back plastic) so he could stick this onto his map. Leaning over the planning table, he unrolled his map, face down onto the fablon, lifted up the map to show everyone the finished product to find that he had fabloned his tie into his sites map. As everyone dissolved in hysterics, Keith picked up a large pair of scissors, cut off his tie, folded his map and sat down ready for the briefing![7]

The two-week deployment covered a weekend with limited flying on Saturday morning, an officers' squadron field dinner on Saturday afternoon and an all-ranks squadron revue in the evening. On my field site, our task was to provide the Harvey

Wallbanger mix for the pre-drinks at the officers' dinner. One evening after flying, we went out on a shower run and bought the Wallbanger ingredients at the local NAAFI. Back at the field site that evening, someone suggested that it would be a good idea to try the mix "just to make sure it was OK". A quick check with the Met man determined that we would be in thick fog in the morning so no chance of flying until late morning at the earliest and so the die was cast for that most dangerous game of 'drinking on the forecast'. After an hilarious evening we all woke up a little the worse for wear to find that, thankfully, the Met man had not let us down and no-one would be flying for some time.

Following breakfast, we retired to the pilots' tent and the coffee machine. Around mid-morning there was still no chance of flying but the site commander suddenly appeared, looked around the sad sight of broken pilots before him, when his eyes alighted on me and he said: "Ah Buswell, you'll do, we've got an aircraft that's just had a water pump change and needs to be checked out in the hover. Get your kit on and come along to Ops." All that was required was to taxi the aircraft out to the landing pad, carry out a vertical take-off, check that the water pump was working then land vertically – job done. The aircraft fuel state had been adjusted to ensure that I would have sufficient power available to hover with the water pump operating although the performance margin in the hover would be very limited if the water pump did not work. (It should be explained that switching the water pump on reset the engine datums to give 107 per cent RPM but, if the water did not flow to provide turbine disk cooling, then temperature limiters would cut in to restrict the RPM – and therefore the power available – to prevent over-heating the engine). As I started up and taxied out it seemed that everyone on the field site came to watch since there was no other activity on site that morning. Checks complete, I was good to go, selected the nozzles to the hover stop, went to full power, checked the RPM rising straight through to 107 per cent and lifted off into the hover. Checking inside the cockpit for the green light which would confirm that the water was flowing I was alarmed to see the light was resolutely not lit. About this time, the 15-second warning appeared to tell me I was on or above the temperature limit for the engine in dry power (since I had no water flow). A quick check inside confirmed that I was sitting on the maximum temperature with the risk of being RPM limited which meant I could start falling out of the hover at any time. The only thing to do (without 'tripping the limiters' which risked burning out the engine) was to freeze on the controls to reduce the bleed from the engine for control purposes and accept where the aircraft went in the hover. I spent the next few minutes wandering idly around the landing pad trying to minimise the control inputs until the engine temperature reduced and I had enough performance margin to attempt a landing which, thankfully, was uneventful. A somewhat chastened pilot, I returned the aircraft to the engineers for further work on the water pump.

I had another memorable moment whilst on this field detachment. During the second week of the exercise when we were being evaluated by NATO (TACEVAL), I was flying No. 2 on a three-ship recce task. The visibility at low level was dreadful but there was pressure to get to the target and get back with the pictures to satisfy the NATO evaluators. Safe in the knowledge that no-one else would be stupid enough to be flying around at 250 feet in such abysmal weather, when we missed the target on the first pass, the leader decided we would go round again to make sure we got some good pictures on a second pass. As we turned back towards the target in a loose trail formation, two Jaguars flew through the middle of our formation at a crossing angle of about 90 degrees at the same height. Luckily the big sky theory worked for us that day.

Back at Wittering in the early summer of 1978, I was programmed to fly to Lossiemouth in the T4. The boss was president of a Board of Inquiry up there so he would fly the aircraft with me 'sand bagging' in the back seat and I would bring the aircraft back to base. The boss had planned a couple of low-level attack profiles in Scotland but I had not been involved in his planning as I had been planning my own trip back to Wittering. It was a nice day and all was going well on the way up north as we approached one of the targets at 250 feet. Cresting over a hill, the boss initiated a ten-degree dive against our target, a bridge at the bottom of a valley. I observed that our line of attack was directly across the valley with steeply rising ground just beyond the target – never a good plan. Closing rapidly on the bridge at 450 knots, I could see that the boss needed to throw the attack away if we were not going to be seriously compromised on the recovery. Unfortunately, this is where 'cockpit gradient' came in: the boss was a very experienced day-fighter ground-attack pilot and I have been on my first operational tour for just 18 months and I am reluctant to say anything. Consequently I sit there and wait for the boss to realise his mistake; at the speed we are travelling this all happens in seconds. By the time I see the boss is fixated on achieving his simulated bomb release, it is too late and we are off the target pulling maximum 'G' to clear the ground rising ahead of us. People who have been in this situation (and got away with it) say that with the nose pointing upwards (so you cannot see ahead of the aircraft) all you get is a sense of the trees and hillside out of the side of the cockpit up around your shoulders. They are correct. We got away with it. We landed at Lossiemouth a little wiser.

TRANSIT RIVOLTO (NORTHERN ITALY) TO LUQA (MALTA) OCTOBER 1978

On my first Harrier tour I was the squadron navigation officer and organised a three-ship 'southern ranger' from Wittering to Malta, night-stopping Dijon southbound, with a lunch and refuelling stop at Rivolto in northern Italy the next day before continuing to Luqa. The other pilots were one of the flight commanders who, in a vain attempt to hide the identity of the guilty I shall refer to as 'Barry', and a fellow junior pilot, Hambone. All was going well on day two and we arrived at Rivolto as planned where, after refuelling the aircraft, we elected to go to the officers' mess for lunch. Entering the dining room we were met by calls of "Hey eeets the Breets" from the Italian aerobatic team, the Frecce Tricolori who were based there, and had also come into the mess for lunch. Space was found at their table and we were made to feel most welcome; they even offered us some wine to go with our pasta – ha ha! (They were having a few glasses so maybe they were going to try something really tricky in their routine that afternoon.) Naturally, Hambone and I declined the kind offer but Barry decided he would partake. In response to our incredulity, Barry explained that when he had been on an exchange tour with the French they all drank wine with their lunch before flying in the afternoon. He had resisted for the first year, moved on to one glass with his lunch in the second year and had gone on to two glasses in his final year in France. The concept of extending this experience to using it after his exchange tour – to adopt the most relaxed military flying regulations of whichever country you were flying through – was an interesting approach, but it seemed to work all right for Barry. After an excellent lunch in such good company, we were all set for an uneventful and relaxed flight down to Malta in beautiful weather (Barry especially) – or so we thought.

Having planned the whole venture, I was well aware that the leg down to Luqa was very tight on fuel. Consequently, as the leader on this leg, I tried to explain to air traffic control (ATC) that we did not wish to start engines until we had a departure clearance and that we would really appreciate an unrestricted climb to our cruising altitude (FL330). Those of you who have dealt with ATC in Italy will appreciate that this was a pretty vain gesture. After a suitable delay which suggested that ATC might, after all, be trying to meet our requests, we were cleared to start up. A few minutes later, our request to taxi was refused while ATC set about trying to get us a clearance. Our mighty Pegasus engines were now consuming vast quantities of fuel even at idle whilst we were getting no closer to Luqa. Some ten minutes later we were cleared to taxi and take off, with a climb restricted to FL140, but at least we were on our way. After 5 minutes at FL140 burning far too much fuel, we were cleared to FL240 where we stayed, still using fuel faster than was good for our prospects of making it to Malta that afternoon.

As leader, I was now in serious fuel calculating mode and desperately trying to put all the range speed/specific fuel consumption/ADD knowledge I could recall into practice. At some point somewhere around the Rome Terminal airspace, my No. 3 appeared alongside me and then pulled up a bit and disappeared backwards. Idiot, I thought, why is he wasting fuel generating overshoot on me only to pull up to drop back into position? I carried on with my deliberations over fuel and diversions.... A few minutes later I looked back right & left to check that Nos. 2 and 3 were with me OK to find that there were no Harriers there. To say the least, this was rather worrying. A quick call to them on the radio brought no response; a call to ATC was met with a deathly silence as well and a call on the standby radio on the emergency frequency of 243MHz elicited no response either. So there I was, somewhere around Rome in controlled airspace, having lost my two wingmen and unable to speak to anyone on either of the aircraft's radios. By now I was searching the skies frantically for some sign of my wingmen. A short while later I spotted two Harriers slightly behind me but at a much higher altitude. So it would appear we had got a climb clearance which I had not heard due to my radio failure, Barry had tried to convey the message when he came alongside me but did not feel he had the spare fuel to hang around going through all the hand signals that might have let me know what was going on. A check of the skies around me, a very quick climb, and I was joined onto the back of my two wingmen. After a while, my main radio came back to life and Barry immediately sloped shoulders and gave the lead back to me. This was a surprise, since my radios were decidedly suspect; on the other hand, it was probably unsurprising in that the decision-making process of where we were going to land was now acute and I had done all the planning – and had not been drinking wine with my lunch.

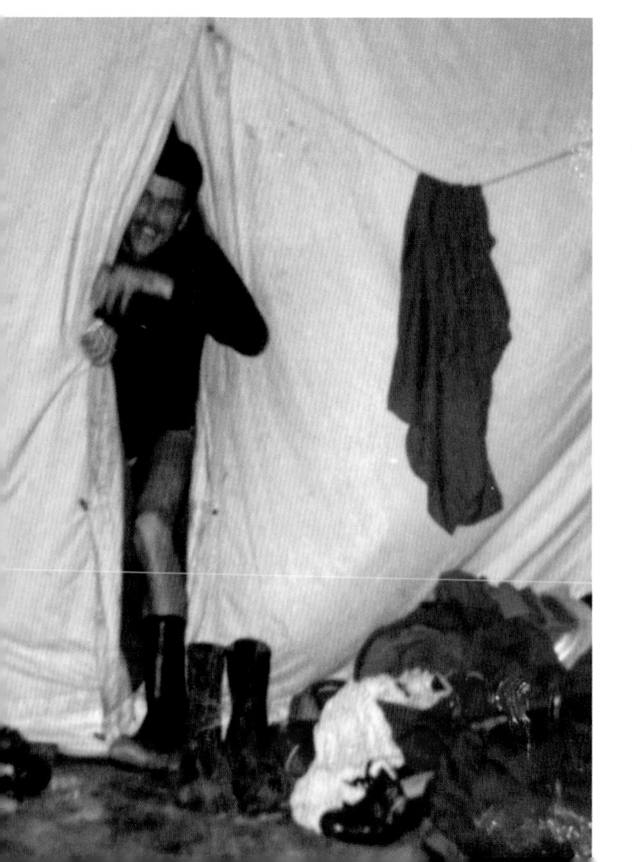

Field living on Exercise Northern Wedding, Kirkwall airfield, September 1978.

Since the weather remained excellent and there were no aircraft based at Luqa any more (the RAF was just about to pull out so the chances of an ATC delay going in there were remote), in true fighter pilot spirit, I elected to press on. As we headed south, it didn't matter how many times I updated the fuel calculations, things still looked very tight. In the beautiful weather we could see the island of Malta a long way out, but that didn't bring it any closer, and despite using the day/night screens to dim out the accusatory fuel Bingo lights, I knew that they were there. Still, I wasn't going to get caught out by the cockpit misting up after a long transit in the cold upper air followed by a rapid descent into warm moist air, and I selected the cockpit demist on in good time (I thought). Descending as late as possible to save fuel we were cleared for a run in and break, most of which was achieved with the throttle at idle to save what little fuel we had left. It was with some relief on my part when the wheels kissed the runway at Luqa but as I increased power for power nozzle braking after landing, the cockpit misted up completely so I could see nothing out of the aircraft. If this had happened just 30 seconds earlier… Fortunately, I was going relatively slowly by then so I opened the canopy, motored the ejection seat to the maximum height and taxied in looking over the top of the canopy arch as though this was the way we always did it.

Once relieved of the Belize commitment, the squadron regrouped at Wittering in the summer of 1978. By this time I had become the squadron QFI, with responsibility for checking out the squadron's pilots, carrying out night currency flights and teaching new pilots on the squadron the art of air-to-air refuelling; I was also a deputy flight commander. We were now under the command of Wing Commander Dicky Duckett, an ex-Lightning pilot who had previously led the Red Arrows. This background gave him a certain amount of kudos which led to a couple of interesting incidents. Shortly after he had taken over command, the squadron was scheduled to deploy on a major NATO exercise (Exercise Northern Wedding) to Kirkwall airfield in the Orkney Islands. Even though the boss was not yet combat ready, he decided that he would lead the squadron to Kirkwall departing Wittering in a diamond nine formation. This was all briefed on the Friday afternoon for the Saturday morning departure. At the end of the brief someone said, "Don't forget it's Burghley Horse Trials, Boss," meaning 'make sure we don't overfly the venue (just north of Wittering airfield), which could alarm the very expensive horses leading to huge claims for damages'. The boss's answer was, "Absolutely – as we come through the overhead of the airfield we'll bend it right to overfly Burghley," which is exactly what we did the next morning and, as far as I know, we never had any complaints. A few days later I was leading the boss and one of the flight commanders on a combat ready work-up sortie for the boss. Halfway through the exercise, a Phantom appeared unannounced so we took him on and after he departed the boss announced he had the lead, told me and the flight commander to join in close formation and then turned back towards the Orkney Islands. As we closed to the coast, we started descending to about 2,000 feet, the speed started building and the boss announced 'pulling up' and he started into his Red Arrows routine with us on the wing looping and rolling.

Throughout my time on the squadron, I was privileged to enjoy the company of an excellent bunch of junior pilots including Bob Marston, Mike Beech, Chris Gowers, Tom Hammond and Clive Loader, who would later rise to the rank of air chief marshal, be knighted and become the first AOC-in-C of Headquarters Air Command. In due course we were joined by Chuck Devlaming, a USAF exchange officer, who certainly added to squadron life in many ways. With a detachment of USAF A6s planned to come to Wittering, Chuck was heavily involved in organising a social programme for their visit. It was decided that a beer call would be accompanied by a piano-smashing competition. This involves teams breaking up pianos with sledgehammers and passing all the broken bits through a toilet seat – first team to get the whole of their piano through their toilet seat wins. Chuck put an advert in the *Rutland and Stamford Mercury* for pianos. Shortly after, he received a phone call from an elderly lady in Stamford who was looking for a good home for her late husband's much-loved piano. Would the piano be going to a good home? "Mam, your husband's piano will certainly be well looked after," Chuck assured her. He was also a pyromaniac and discovered a miniature brass cannon on the squadron which he thought he could get to work. At the next squadron dining-in night, there was suddenly a cry of "fire in the hole", followed by a loud bang as the cannon went off and trouts' eyes, collected from

End of tour on No. 1(F) Squadron. From left: RN Midshipman (holding officer), Chris Gowers, Bill Tyndall, Sqn Ldr Barry Horton, Tom Hammond, Wg Cdr Dickie Duckett (boss), Author, Sqn Ldr John Finlayson, Clive Loader, Maj John Mosely (army GLO), Chris Bain and Mike Beech.

the starter course, were scattered across the mess dining room. The follow-up to this was that at a summer barbecue in the mess garden, Chuck's cry of "fire in the hole" was followed by an almighty bang as the cannon (which had not been exposed to any serious NDT) exploded with bits of brass flying around the garden. Miraculously no one was injured.

Much to my chagrin, the postings branch in Gloucester had now decided that I was to be short-toured on 1(F) Squadron to move back to the operational conversion unit as an instructor. This was not my plan at all as I was expecting to move to one of the Germany squadrons the following year. With a number of Harrier mates in the Red Arrows, and having often thought about a move to the team, I immediately put in an application as a way to circumvent this posting. To no avail – the postings branch rejected my application as there was no-one else qualified to take my place at the OCU but told me that they would look favourably on my next posting (I was also interested in an exchange tour) when it was possible to replace me as an instructor. So in November 1978, I very reluctantly moved back up the airfield to 233 OCU. At my dining out from the squadron, along with the other officers leaving at the time, I was subjected to the customary squadron farewell – a verse composed by my fellow officers and sung to the tune of 'Bye Bye Blackbird':

Chris Burwell QFI
Settled in the bachelor life
Bye Bye Brunswick
Got his pad and his fast sports car
Cursed thing won't go so far
Bye Bye Brunswick
Getting into ho-les is his hobby
So OCU should really make him happy
His tour on One is left undone
We're sorry shag cos you've been fun
Brunswick Bye Bye

CHAPTER 6
HARRIER INSTRUCTOR AND DISPLAY PILOT 1978–81

After the excitement of two years on 1(F) Squadron and professional development as a front-line offensive support pilot and squadron QFI, a tour on the OCU was bound to be a bit of an anti-climax. Despite the assurances of the postings branch, and notwithstanding my best efforts to influence it otherwise, this tour was almost three years long. However, instructing the basic VSTOL exercises from the back seat of the Harrier T4 was a very demanding job. One of the key attributes of any flying instruction is knowing how far to let the student pilot go before stepping in and taking control. In a Harrier, where the pilot has a nozzle control that will instantly change the direction of the 20,000 lbs of thrust of the Pegasus engine through more than 90 degrees, this judgement can be critical. It was not unusual, when things started to go wrong, for the niceties of "I have control"/"You have control" to be overtaken by a strangled cry of "I've got it" from the back seat as the instructor was required to salvage the impending crash. It is a tribute to the standard of instruction on the OCU and the quality of the students we had that training incidents (or worse) were few and far between.

During my time instructing we trained all the early Sea Harrier pilots who would subsequently go on to fly in the Falklands War. I got on well with them and after 18 months' instructing volunteered to move to Yeovilton on loan to the Fleet Air Arm along with a few other RAF pilots who were being sent there to bolster the RN squadrons' experience levels. Once again, I was told that I could not be replaced at the OCU. Around this time one of my fellow instructors, Tim Allen, who would go on to become a military then a civilian test pilot, started to explore the operating envelope of the Harrier. We all learnt a lot from Tim's exploits about the capabilities of the Harrier; not least, John Farley, the renowned BAe Systems Harrier test pilot, visited Wittering to find out what Tim was doing with the aircraft.

In 1979 I volunteered to take part in another RAFG TACEVAL, this time with No. 3(F) Squadron, and I was delighted to be allocated to the same field site as my good friend Paul Hopkins from training days at Cranwell and Valley. However, this field deployment took a more dramatic turn than the last one with IV(AC) Squadron.

PARTING COMPANY 1979

This account was originally published in *Harrier Boys Volume One* by Bob Marston.

Last night we went to a bar. Pilots go to bars not just to drink beer, but to tell stories – flying stories. Last night Hopkins told a story. Inevitably, being Hopkins, that meant he had to accompany the story with all the actions. As it was a Harrier story, it involved a lot of 'left hand' with the throttle and nozzle lever. At the moment critique, as his hand shot forward to select full power, it inadvertently propelled a full glass of beer over me and my only flying suit. The stories were good and the beer was good. As we drove back to our dispersed Harrier site and the camp beds in our tents, the smell of my beer-soaked flying suit was not so good. That was last night.

It is now morning. It is early autumn. The weather is fine and the dew disperses slowly in the warming sun. We are operating our Harrier short take-off and vertical landing aircraft from a remote field site in West Germany. Lez and I have now been in our two aircraft for some hours and we have flown two offensive support missions as a pair. When we land we stay in the aircraft and we are tasked for our next 30-minute sortie on a landline. This time there is a delay and we sit tight, bantering on the landline. The frivolity is a welcome release from the mind-focusing reality of taking 20,000 lbs of Her Majesty's fighter aircraft off from a tiny field strip, with trees and electricity pylons at the end of it, and delivering it back safely, vertically, half an hour later. It is hot in the cockpit. Eventually, we are tasked. Our ground liaison officer (an army major) tells us that we have to go to a certain position and contact 'Fortune' (a forward air controller) on a given radio frequency. He will direct us to our target. Time is now of the essence if we are to make our 'time on target'. After planning and briefing the mission in our cockpits, we speak quickly with the 'authorising officer' whose last input is to make sure that our ejection seat pins are removed from our seats. This is to ensure that we can eject from the aircraft in emergency whilst airborne should anything catastrophic happen.

Engine started, in radio silence I taxi out across the grass ahead of Lez's aircraft, line up on the ridiculously short strip, keeping rolling as I check flaps, armament masters, short take-off stop set in the nozzle quadrant, trims, select water injection on and hit full power. This aircraft is outrageous. The acceleration is staggering. The combination of acceleration and vibration as the aircraft bucks across the grass strip, ever nearer towards the trees and pylons, means it is almost impossible to read the instruments. The speed is racing through 90 knots. A flag in the ground tells me at which point I must move the nozzles to the short take-off stop. Instantly we are airborne; the jarring, bucking, careering ride across the grass is replaced by flight, smooth flight, as man and machine enter an upward trajectory, but not one that will clear the trees and pylons. Without delay

I must now rotate the aircraft to its maximum angle of climb to clear the trees and the electricity pylons. Once clear of the obstacles, I must then lower the nose of the aircraft, accelerate, raise the undercarriage and flap, reduce power, switch off water injection – and remember to breathe. I have been flying this aircraft for over three years and it is my life, my *raison d'être*. What else would I be doing in life if I were not flying a Harrier GR3 of Her Majesty's Royal Air Force?

For some strange reason, I am leading this mission. Strange because I am not based in Germany but have come out from the UK to take part in this exercise. Lez is based in Germany and he should know his way round Westphalia better than me; especially on a day like today when, although it is now fine and warm, the visibility is 'not good'. Nevertheless, we meet up after take-off and set off to call Fortune on the radio, get our airborne brief and find the target. It's another routine day in the office, travelling through the north German countryside at seven miles a minute in five kilometres visibility at 250 feet, evading simulated Soviet MiG fighters and surface-to-air missile systems.

I have known Lez for some years and we enjoy each other's company. Flying, especially military flying, is about camaraderie and Lez and I will be good friends for a number of years to come. Already we have trained on the Harrier at the same time and we have flown together on peacekeeping missions in Belize, Central America; in the future we will operate together over a long period in Germany and for some months in the Falkland Islands after the war. This means we will share our lives; our families will become friends and we will fly and drink together and enjoy the support and mutual respect we have for each other. Yet, like all military fliers, we will also play down the ridiculous reality of what we do for a living. If the political balance between East and West loses its poise, Lez and I and many front-line aircrew across all the NATO air forces will be at the forefront of hostilities. It is doubtful that we would survive for very long. But that is for tomorrow; today we have to go and speak to a man called Fortune. The visibility gets no better, but Fortune is there and he gives us our airborne brief.

As we prepare for the attack run at 450 knots there is a muffled thump from behind me in the aircraft. This is not good; this is where the engine sits. I say engine advisedly because there is only one. If it stops, it is not possible to pull over to the side of the road and lift the bonnet. I check the engine speed against jet pipe temperature, I check the two hydraulic systems, I check the central warning panel which should tell me if my aircraft is not well; all the indications are normal. I ask Lez to come and have a look at my airframe to check for any signs of damage. I call my speed at 350 knots and I have now climbed to 1,500 feet and I turn towards our main base airfield. Lez comes alongside and assures me all is well. Ironically, at this point the engine fails completely with a deep rumble, heavy vibration and a total loss of thrust. With no power from the engine, the only way to keep airspeed, or airflow over the wings, is to lower the nose of the aircraft into a glide. Unfortunately this means that I have now made an imminent appointment with the ground, unless I can persuade the engine to work again. I am now looking at a lot of ground coming up to meet me in a hurry. A Harrier has the gliding characteristics of a brick.

Pilots do not like simulators. Pilots like to be master of their aeroplanes, to be fully in control; they are magicians in three dimensions and they have a flair and panache that mere mortals do not possess. But pilots are stretched to the limits of their ability by simulators and the devious machinations of simulator instructors and sometimes, in the simulator, their flair and panache abandons them. Today I am glad that I have done countless engine failure drills in the simulator. I shut down the fuel flow to the engine, I select flap, I select the manual fuel system and I try to relight the engine. I do everything right and the engine relights. But it will not produce any thrust. I still have height in hand so I go through the drill once again but with no more success. I am now quite low but there is just time to point the aircraft towards a large wood, shut the engine down for the final time, call "ejecting" on the radio, sit up straight in the ejection seat, head back, take a deep breath and pull the ejection handle. My last thought is that it won't be too good for me if this seat does not work as advertised. But I have every faith that it will.

The vertical acceleration imparted by the ejection seat is hard – very, very hard. Both physically and psychologically it is shocking in its violence. It is as much, and sometimes more (as I shall find out in a few years' time) than the human body can withstand. I have a fleeting glimpse of my lifeless, doomed aircraft rushing away from me as I hurtle vertically upwards away from impending disaster. Within one second, with a sudden and extreme jolt, my parachute deploys and the world around me becomes ethereal. Peace. Silence. Warm country air. Green fields. I am alive – and it is good. From a frantic, rapidly unfolding nightmare, I have been catapulted into a place where I am suspended 500 feet above the ground, swinging gently in the parachute with no sense of descending at all. It would be good to stay here all day. But now training and reality return as I can see that I am in fact descending and I must prepare for the landing, my first ever by parachute. I get rid of my oxygen mask, lower my personal survival pack so that it hangs well below me and try to assess which way the wind is drifting me. The arrival is firm and accompanied by a severe pain in the ankle; it must be broken, or at least very badly sprained. But I am still alive and will live to fly another day. The exhilaration of being alive, of smelling the grass on which I am lying and of looking up into the hazy blue sky is unspeakably intense. All around, the birds complain about my violation of this glorious morning. To be able to hear them is a joy, but how can I make amends?

I have parted company with my Harrier over sparsely populated farming country yet within minutes of landing I am surrounded by German people, all very concerned for my welfare. The blood on my face exacerbates their concern. This has been caused by cuts in my cheeks from the molten metal detonating cord used to shatter the canopy on ejection.

> These bits of metal have embedded themselves in my face and will still be coming out of my cheeks for some years to come. Cuts to the face and a damaged ankle are a small price to pay for being alive and unexpectedly sharing my morning with the birds.
>
> With impressive timing, one of our helicopters appears from nowhere and takes me to our base airfield for a medical check before transfer to hospital. As I strip down, I receive quizzical looks from the doctor and nurses as I remove and hand over a very dirty and rather sweaty flying suit. It has a distinct smell of stale beer about it.

Returning to the UK, I was flying again within three weeks having had the plaster removed from my leg (to allow a badly sprained ankle to recover) and no-one any the wiser that I had a small compression fracture in my neck. The problem with my neck did not manifest itself until some years later, although it was a very small price to pay for having my life saved by Martin-Baker. As far as I am concerned, no praise is high enough for this company that makes ejection seats for military aircraft across the world. Sitting on the ejection seat every time we flew, it was, perhaps, too easy for us to take their product for granted but it was at times like this that we were reminded what an outstanding product (and lifeline) they provided us with. As recalled above, I can still remember that momentary thought, just before I pulled the handle, that I had 'every faith' that the seat would work as advertised.

Training fast-jet pilots is not without its risks. During my time on the OCU I had two close calls both of which, I am embarrassed to admit, were ultimately my own fault. Carrying out a 1 v 1 air combat training exercise one day, I was in the back of a T4 with a student in the front who had just returned from two years' loan service in Oman flying the Hunter. Operating against one of my fellow instructors in a GR3, we got into a nose-high scissors situation where whoever manages to fly slowest will end up behind his adversary. Since this is a potentially risky situation with both aircraft crossing close to each other with reducing speed (and therefore controllability), rules are laid down as to which aircraft avoids the other to avoid a mid-air collision. As the

Post-ejection, back at the field site, October 1979.

Presentation of ejection seat handle by Les Hendry – officer in charge of the Ejection Seat Bay – at RAF Gütersloh.

speed reduced and we entered into a potential conflict, it was obvious (to me) that we would have to give way by lowering the nose and going under our opponent. Lulled into a false sense of his abilities due to my perception of his flying experience in Oman, I waited too long before realising that my student was not going to do this and had set up a collision course with the other aircraft. Too late, I took control, rolled the aircraft away from the GR3 with what little control we had left (we had none in pitch as we were essentially ballistic), and put the nozzles to the hover stop to try and reduce our closure on the GR3. All I remember then was curling up in the ejection seat and waiting for the impact; luckily this didn't come. I called the fight off, lessons learnt, and we all lived to fly another day. Very sadly, a few years later, a 1v1 OCU combat did end in a mid-air collision in which the instructor and student in the T4 were killed; the instructor in the GR3 ejected and survived.

On another occasion, having just qualified as an instrument rating examiner, I took a refresher student up on his instrument rating test – my first instrument rating test as an examiner. With his extensive previous fast-jet and Harrier experience, I was expecting it to be a straightforward first test for me and he was expecting to be awarded a master green rating. It quickly became apparent that it wasn't going to be as easy as I thought and the refresher student was having some trouble flying to a good standard. The test profile required a practice diversion to another airfield followed by steep turns on instruments at 2,000 feet, using 60 degrees angle of bank. As we departed the diversion airfield, the examinee was having a lot of trouble settling the aircraft level at 2,000 feet. I advised him not to rush as entering steep turns when unsteady in height was only going to lead to an unsettled steep turn. As we headed south with a layer of cloud beneath us, the cloud tops started to come up towards us at 2,000 feet and I really needed to stay out of cloud for the steep turns as we were not under radar control. Eventually he settled the aircraft down and we managed to start the first turn but halfway through the turn we began to enter cloud.

Due to the delay starting the turns and the high recovery fuel state at base owing to the weather, we were now getting tight on fuel to complete the test profile. I therefore made the decision to continue the turn and get back into clear air rather that stop the turn and reposition for another attempt. (We were over the Fens so ground and obstructions were not factors.) After a short while in cloud, we got into the clear, completed the rest of the profile and returned to base. I was not impressed with myself over my lack of airmanship, but consoled myself with the fact that there was little flying taking place due to the poor weather and the risk of anyone else being in that bit of cloud at the same time as us was hardly worth thinking about. Relaxing in the bar with a beer at the end of the day, I got talking to a friend from air traffic control who told me about controlling a Harrier on a radar approach that morning when another aircraft suddenly appeared on the radar screen co-altitude with his traffic and on a collision course. He did not have time to give his aircraft a turn to avoid this traffic and thought they were going to collide. We didn't. I left the bar feeling very small.

One other bit of excitement came one day when I was flying in a T4 with one of my fellow instructors, Geoff Timms. I was the aircraft captain but Geoff was flying the aircraft shortly before we returned to base. We had some fuel in hand and Geoff wanted to practise some 'viffing'[8] before landing. Viffing can be used at higher speeds to increase the rate of turn for a short time until this is offset by the decay in speed; it can also be used to maintain control down to very low speeds but this can lead to a loss of control if not exercised very precisely. This loss of control is notably marked by a 'flick' (aka a departure, an auto-rotation or an incipient spin) which, if allowed to continue, will lead to a fully developed spin. We had all experienced many departures but most of us had not managed to enter a fully developed spin. Fully developed spins are not permitted in the Harrier due to the huge loss of height incurred in the spin and in the recovery. Consequently, the minimum ejection height for a fully developed spin in the Harrier is 10,000 feet. As he practised his viffing on this day, Geoff managed to flick the T4 off a low-speed manoeuvre with the nozzles deflected at about 9,000 feet. This was not a problem as the Harrier was very consistent in recovering from a departure straightaway provided the nozzles were put fully aft (normal flight position) and the flying controls centralised. Geoff did all this correctly but the autorotation continued. This woke me up with a start as a glance at the altimeter showed that we were at 8,000 feet and below safe ejection height for a full spin. As the aircraft nose dropped into what I thought was going to be a fully developed spin (and I would probably have to order an ejection even though we might not survive it) the autorotation stopped and we recovered at about 4,500 feet. This was the only time that I experienced the Harrier not recovering immediately from a departure once the controls had been centralised and the nozzles run aft.

Towards the end of my tour on the OCU, one of our recent graduates, Nigel Storah, was killed in an accident at Gütersloh in Germany. Approaching the hover, the aircraft rolled rapidly out of control; Nigel ejected, sadly too late, and was killed on impact with the ground[9]. Although it was quite feasible to lose control of an aircraft in roll in VSTOL flight if the aircraft was allowed to develop sideslip, this did not appear to have been the case here; not least, Nigel was a capable if inexperienced Harrier pilot, an ex-creamie, and such an error would have been uncharacteristic. With this kind of unexplained accident, it was expected that the Harrier would be grounded pending, at least, the initial accident investigation. This did not happen and we continued flying. A day or two later, Syd Morris experienced a loss of control in roll as he commenced a deceleration to land. Fortunately, he still had aerodynamic control and accelerated back into fully wingborne flight and landed the aircraft conventionally (i.e. without the use of vectored thrust). After landing it was discovered that the control rod operating one of the jet reaction controls in the wingtip that controlled the aircraft in roll in the VSTOL regime had broken. Referring back to Nigel's aircraft, it was determined that the same thing had happened to his aircraft. At this point, all Harriers were, somewhat belatedly, grounded and only returned to flying status once the roll control rods had been replaced. As the flight safety officer on the OCU I wrote to the inspector of Flight Safety through the confidential reporting system, questioning why the aircraft had not been grounded following the fatal accident. The decision to keep flying, even though the reason for Nigel's accident was totally unknown, had almost led to the loss of another Harrier (and potentially the pilot). I did get a reply citing the need to maintain operational credibility (for NATO or the Warsaw Pact?) which, to my mind, did not excuse the risk to the pilots flying the aircraft at that time.

As my nominal two-and-a-half-year tour was coming to an end I put in another request to join the Red Arrows. I couldn't see how they could stop my application this time but it now transpired that I was to be promoted to squadron leader in July and my application was indeed rejected by the personnel branch once again. I had also asked if I could do the Harrier displays that summer and at least this was approved. Dave Linney and I both trained up for this task in the early part of the year. It was agreed that we would share the displays between us with the other bringing the spare aircraft to the displays. Having worked out the display profile we would use, we had to practise one particular manoeuvre at a safe height before bringing it down to low level. This manoeuvre was an acceleration from the hover into wingborne flight, a pull-up into a half loop with a roll-off-the-top into a steep decelerating transition back into the hover at the original starting point. It had been thought up by one of

The author as a display pilot at Staverton in May 1981. His nephew Simon is in the cockpit whilst niece Clare is stood in front of the aircraft.

my fellow instructors, Dudley Carvell, a very experienced Harrier display pilot (and ex-Red Arrow) who had kindly stepped aside to allow me to display that year.

Whilst the majority of the display did not stray much beyond the capabilities of most reasonably experienced Harrier pilots, this manoeuvre was something we certainly didn't normally do and needed to be well practised and thought through for every display; it was also the one that caught me out in training one day. On one of my first low-level practices of this manoeuvre, all appeared to go well and I arrived back in the hover as planned. At this point, a radio call came for me to 'check gear'. I looked inside the cockpit to see that I had arrived back in the hover with the undercarriage retracted. Having correctly retracted the gear accelerating away from the hover, I had been so pre-occupied with rolling off the top of the loop and commencing the steep decelerating transition that I had omitted to lower the undercarriage. Full marks to Officer Commanding 233 OCU, Wing Commander Peter Millar, who was supervising me – no radio call was made until I was back in the hover since to call for me to check the gear whilst I was carrying out the steep decelerating transition could have been a dangerous distraction. Undercarriage lowered, I then continued with the practice, no damage done apart from to my pride, and a valuable lesson learnt. Having got our display clearances from the AOC, Dave and I had an enjoyable, if limited, display season due to fuel cutbacks across the RAF.

My most memorable display was in France, operating from a small club airfield, where I experienced an undercarriage problem right at the end of the display just before I was about to land. Being aware that the airfield I was at had no hangarage or support facilities for what I knew would be some extensive work to repair the aircraft, I accelerated away from the hover with very little fuel remaining and diverted to the nearby military base. Two days later an RAF C-130 Hercules arrived with a hydraulic rig and a team of engineers from Wittering to sort out the aircraft, which I eventually flew home after an extended stay in Bordeaux. I was stuck in France over 1 July which was my promotion date to squadron leader and consequently met the Hercules crew still wearing flight lieutenant rank slides. The Hercules captain gave me an envelope containing a note from my boss, Squadron Leader Bill Green[10], congratulating me on my promotion and enclosing some squadron leader rank slides which I put on when no-one was around. This confused the C-130 crew completely, since I had suddenly and quite unexpectedly become a 'senior officer'.

Denied the opportunity to join the Red Arrows once again, I now had a posting to No. 3(F) Squadron at RAF Gütersloh in Germany as a flight commander.

CHAPTER 7
NO. 3 (FIGHTER) SQUADRON, THE CENTRAL REGION AND THE FALKLAND ISLANDS 1981–84

I joined the squadron in September in time for the final field exercise of the year under-studying John West as site commander. Whilst my focus should have been on reaching combat ready status as quickly as possible, no sooner had we got back to base than I found myself nominated to run a Board of Inquiry (BoI) into a Jaguar accident at RAF Brüggen. The aircraft had suffered an engine run-down following a lightning strike and the second engine had failed shortly before landing at Brüggen resulting in the pilot ejecting and the aircraft being a write-off (Cat 5). The RAF had a system of aircraft accident Boards of Inquiry being run by a senior aircrew officer (more senior to the captain), supported by specialist members as required. As I was not a Jaguar pilot, I had a Jaguar pilot and a Jaguar engineer nominated to assist me.

We very quickly ascertained what we thought had happened although the pilot was adamant that he had followed the engine failure drills correctly. Consequently (and this was one of the advantages of the peer BoI system for the pilot involved), we then spent many weeks trying to work out another explanation for the cause of the accident. This involved various trips to the UK, most notably a very interesting trip to the Air Accident Investigation Branch (AAIB) at RAE Farnborough. Somewhat disturbingly, this visit included a tour of their hangar which contained the wreckage of many of the recent UK air accidents which were under investigation at the time, many of which were fatalities and some involved people we knew. Our visit to the AAIB effectively confirmed our conclusions so we subsequently asked one of the RAF's aviation psychologists to visit us in Germany, interview the pilot and help us produce an explanation of the disconnect between our view and the pilot's being in denial about getting the engine shutdown drill wrong.

At this time, the Central Region in Germany was the place to be: at Gütersloh we were facing the Cold War Soviet threat which was just 70 miles away from our base. In the event of hostilities, our Harrier concept of operations was to deploy off the main base to pre-nominated sites which had been recce'd but were a closely guarded secret.

On practice exercises we would deploy six flying sites, with six aircraft on each site, a HQ and a number of logistics parks providing all the support including fuel, weapons and food. We departed on our missions from a short strip – it could be a road, metal planking or grass – and recovered back into the site vertically onto a landing pad laid by a Royal Engineer detachment. As we did not take our usual 100-gallon drop tanks on the aircraft into the field (to allow for the limited take-off distance available and to preclude the error of trying to take off with too much fuel on board[11]), our normal sortie length was just 30 minutes. It was then quite usual to stay in the cockpit to carry out a cockpit turn-round then fly another mission. The cycle could be repeated with pilots staying in the cockpit for up to four missions with the possibility of returning to the cockpit later to complete a maximum of six sorties per day. This was a very high workload on the pilots in a high-risk operation with little scope for error. To add to the pressure, when the air temperature was high (which it often was during the two summer field deployments), it was quite normal to approach the vertical landing pad still above dry hover weight (i.e. relying on the water injection system to get into the hover). By the time the water ran out (normally after about 70 seconds having used some water for take-off), you had got rid of 500 lbs weight of water and burnt enough fuel in the approach to hover without water injection – or so the theory went. If you had a problem coming into the field site, you might just have enough fuel to divert to the nearest airfield. Since the field sites were well camouflaged, it was not unknown for one of the inexperienced pilots to end up in the hover, running out of fuel very fast, but without the landing pad in sight. In such a case, he would break radio silence and ask if anyone could tell him which way to go to get to a pad. Occasionally inexperienced pilots managed to land at the wrong field site which they weren't allowed to forget in a hurry.

Very early on in my tour I had to deal with a problem with one of the pilots on my flight. He had previously come to me towards the end of my tour on the OCU at Wittering whilst he had been going through the course, to say that he was not convinced that he wanted to continue training on the Harrier. It appeared that, quite understandably, his concern was based on some level of under confidence. I assured him that he was doing fine and I could not see any real problem with his performance; he had subsequently continued with his Harrier training and had been posted to 3(F) Squadron and to my flight in Germany. In January 1982, he came to me whilst we were on armament practice camp for three weeks in Sardinia to say that he had experienced what would appear to be some form of depersonalisation disorder, where he felt outside his body, when flying at high level down to Sardinia. This was a known phenomenon amongst pilots. Without taking the matter further, although I think I did discuss it with the boss, I arranged to fly the T4 back to Germany at the end of the detachment with the pilot in the front seat. This we did with no untoward experience and he went on to have a successful career, ultimately commanding a Harrier squadron and progressing to higher rank.

Our first major deployment in 1982 was to Exercise Maple Flag at the Royal Canadian Air Base at Cold Lake in western Canada. With a clearance for low-level operations down to 100 ft for this exercise (as opposed to the normal peacetime limit of 250 ft), we detached to RAF Lossiemouth in the Moray Firth for two weeks to work up (or down) to 100 ft flying. After initial singleton sorties, where most of the attention was out of the front of the aircraft to maintain a safe clearance from the terrain and obstacles (small masts and some pylons are not easily seen when you're approaching them at seven to eight miles per minute), by the end of two weeks we were all quite comfortable flying in tactical formations of four or six aircraft with the ability to keep a good visual check in all sectors. For this training we used Jaguars from 226 OCU and the Hunters from the Buccaneer Wing as offensive fighters.

As the detachment to Canada approached, and with the Falkland Islands becoming an issue, 1(F) Squadron who were air-to-air refuelling qualified flew the aircraft across the Atlantic to Goose Bay and we took over the aircraft there to transit them across the northern wastes of Canada to Cold Lake (three long legs using the 330-gallon drop tanks for extended range). With six aircraft, the boss led the first leg which didn't go too well as his speed control in the cruise at high level was erratic – he'd either had a bad night or his machmeter was giving problems. I took over the lead for the remaining two legs. Departing on the next leg, there was cloud at about 20,000 feet and, since fuel was tight, I could not afford to spend too long at lower levels getting everyone together before going into cloud. By the time I reached the thin cloud we still had two stragglers who took separation from us and continued the climb on a parallel track. (In the northern Canadian wastes there was no radar coverage so everything was done on a procedural service). After some time, No. 5 called to say that they had our contrails and were closing up on us. I thought this was rather odd as they should have been well out to our side and should not have been able to see our contrails. A short while later we came out of cloud and I looked out to the right to see an airliner being stalked by two Harriers! They slid across to join us, no more said.

The exercise was as close to going to war as most of us had been with the number of aircraft taking part, huge tactical freedom and a capable and aggressive air threat (inter alia the USAF aggressors in their F-5s). However, at the debrief on the second day, the aggressors said that they had not seen a single Harrier and were out to find us the next day. Given the height we were flying at, the small size of our aircraft and very effective camouflage, this was not a great surprise. The aggressors worked out the only way they were going to find us was to get down in the weeds and this they did with one of their aircraft actually going underneath a Harrier on one mission! Of course, operating at such heights increased risk and one of our Harriers did make contact with some trees although, given that no damage was done, this was kept quiet. On one mission, our formation was split up by fighter aircraft and I met up with another Harrier as I ran into the target area. He joined me for the pre-briefed pairs attack and as I approached the target with the gun camera running to record my bombing attack, I saw something interesting close to the target. After landing I walked back in from the aircraft with Mike Allen, who had been on my wing for the attack, and asked him if he had seen anything in the target area on our attack? No was the response (he would have been busy hanging onto me), so I suggested we have a look at my film. Sure enough, as we ran into the target area right at the back of our TOT (time on target) bracket, some USAF Air National Guard F-105s had started dropping practice bombs onto the same target right at the start of their TOT bracket. It was a demanding two weeks; as the squadron QFI, Jonathan Baynton, said to me at the end of the exercise, at least we all came out of it alive.

With all the RAF's air-to-air refuelling tankers committed to supporting the Falklands build-up, we now had the task of returning the aircraft to Gütersloh without their assistance. I took on responsibility for planning the routeing of our Harriers back across Canada (three legs in one day using the 330-gallon drop tanks: Cold Lake–Winnipeg; Winnipeg–Bagotville; Bagotville–Goose Bay). After a day at Goose Bay for the engineers to sort out any problems with the aircraft, we planned Goose Bay–Søndre Strømfjord (Greenland); Søndre Strømfjord–Keflavik (Iceland); and Keflavik–Lossiemouth (Scotland) for another day to cross the Atlantic. We lost our Nimrod top cover (to provide us with additional sea survival gear should we eject) in the afternoon and wisely spent the night in Iceland before continuing to Lossiemouth and Gütersloh on the final day.

Back in Germany, life was very focused on events in the South Atlantic. Some of the ex-1(F) Squadron Harrier pilots were being loaned to the Fleet Air Arm with their AAR experience, as were two ex-Lightning pilots who were now on Harriers – John Leeming and Steve Brown – since they had single-seat radar experience. As an ex-1(F) Squadron pilot, I was expecting I could be called upon but given my flight commander role in Germany, I thought this might not happen. In the event, I found out from my boss, Wing Commander Bob Holiday, that he had blocked my move from the squadron in Germany, not least because my wife was about to give birth to our first child. One of my abiding memories of this period was waking up every morning and listening to the news from the Falklands which often included details of the loss of one or more Sea Harriers. Later in the day we would get the names of which pilots had ejected and who had been killed, all of whom I knew from my time instructing them on the OCU at Wittering. I had mixed feelings about not being present in the South Atlantic. Like most others I believe, my own view on getting involved in hostile action, which would invariably result in the taking of life, was that it was fully justified once diplomacy had failed. There was no doubt in my mind that this situation pertained in the Falklands. The Harrier Force having been committed to action, it was frustrating from a professional point of view not to be involved. There was also a feeling of some guilt that others were being asked to lay their lives on the line whilst we remained relatively safe in Germany. And, of course, a number of the Sea Harrier guys didn't come home. However, I cannot deny that from a family perspective it was obviously about the worst time possible to go to war.

The RAF Harriers and the Sea Harriers acquitted themselves well during the conflict although the relationship between the RAF and the 'fish heads' driving the carriers was not good, as candidly recounted by Peter Squire, OC No. 1(F) Squadron[12]. Back in Germany we had the task of training up and providing reinforcement pilots for the Falklands theatre. This entailed a number of detachments to Lossiemouth where we could maintain our 100 ft flying currency and use the air-to-ground weapons range at Tain. As the Training Flight commander, organising these detachments fell to me but the support from HQ RAF Germany was excellent. Late one week I phoned Squadron Leader Roger Gault at HQ and said we needed some fighter support on the Monday for our week at Lossiemouth. When we arrived in Scotland on Monday morning, there were six Lightnings from Binbrook waiting for us – our fighters for the week – with hand-picked pilots from the Training Flight, and 5 and 11 Squadrons to give us the best training possible.

It was also decided that all Harrier pilots should now be qualified in air-to-air refuelling which, until now, had been the preserve of 1(F) Squadron. Given the imminence of potential detachments to the Falklands and the difficulty of arranging Harrier T4s and tanker aircraft for training, I decided that I would take a team of pilots to RAF Marham in Norfolk to complete their tanker training in single-seat aircraft. Having led a six-ship of GR3s into Marham the day before, we attended tanker ground school in the morning and I briefed the five pilots very thoroughly on the techniques to employ and the safety aspects. We then flew into the North Sea where we joined our Victor tanker, I demonstrated a 'plug in' (the first time I had tanked for a few years) and then they all had a go. Much to my surprise and relief, there were no problems with any of them and they were all now 'day tanker qualified' – no time for any consolidation. Landing off that sortie into Yeovilton, I was briefed on ski-jump operations by one of the Sea Harrier pilots that afternoon, did three launches off the fixed ski jump (replicating the inclined ramp on the front of the aircraft carriers) and flew back to Germany late in the afternoon – air-to-air current and ski jump qualified! In the event it was not until early 1983 that I eventually detached to RAF Stanley in the Falkland Islands for the first time.

THE FALKLAND ISLANDS FEBRUARY–MAY 1983

In the aftermath of the Falklands War in 1982, the RAF set up an 'air bridge' from Ascension Island to RAF Stanley using the C-130 Hercules transport aircraft with air-to-air refuelling support from Victor tankers. This was a very impressive operation which entailed the C-130s being refuelled as they headed south from Ascension giving them enough fuel, if they were unable to land at Stanley (typically due to the weather), to return to Ascension – a flight in excess of 20 hours with only one crew. Since the resupply to the Falkland Islands from the C-130s was limited compared with the total requirement, much of the resupply was carried out by sea and that is how I came to be on a cruise ship, the *Cunard Countess*, in February 1983 en route from Ascension to Stanley in charge of a detachment from 3(F) Squadron. With me I had three other pilots from 3(F) Squadron and a team of about 30 of our engineers. The ten-day cruise south was very relaxing. We organised two sessions of PE training for everyone every day led by one of my pilots, Dez Dezonie; in addition, there was training for promotion exams and I was undertaking a correspondence course of staff training. There was still plenty of time to swim in the pool and socialise with the other passengers on board. A week into the trip the weather deteriorated as we headed further south from the equator. The seas started to build, it got much colder and the pool was drained. The final round of cocktail parties was undertaken on pitching decks and, on the final morning, as the Falkland Islands eventually came into view, it was a somewhat muted

Morning physical exercise on the *Cunard Countess* en route to the Falkland Islands, February 1983.

gathering on the windswept desk that stared out at the desolate islands that were to be our home for the next three months. More pilots from the squadron joined us at Stanley via the air bridge and we took over the Harrier detachment from IV(AC) squadron in mid-February.

On arrival at the Harrier Flight at RAF Stanley, we quickly took over from our predecessors and settled in for our three-month tour of duty. I could see that maintaining a disciplined focus, both professionally and socially, was going to be a vital issue, especially given the small size of our detachment and the limited work especially for the pilots. On our first day running the detachment, the first crews back from flying in the afternoon grabbed an early tea and settled in to watch a movie without a thought that the rest of us still out flying might want to watch the film as well. Once the film was finished, I called a meeting and suggested that in future we might want to start a film at 1900 so we could all watch it together. Having already seen the Phantom detachment sitting around all day watching films, I further suggested that we would only watch films in the evening – nothing before 1900 – and this was accepted and adhered to throughout the detachment. It was interesting to see our successors arrive three months later, drop their bags in the crew room tent, put a film in the VCR and sit down to watch it – no thought of sorting themselves out or getting organised – an interesting approach to arriving in an operational theatre. I was happy to leave it to them to sort how they were going to manage their three months in the South Atlantic.

At the end of an 8,000-mile supply line, aviation fuel was a precious commodity so our flying was limited although we did have a standby commitment: in the event of any Argentinian aggression or provocation, we were required to be at 15-minute readiness throughout daylight hours. Since the Harrier GR3 had no operational night capability, the air defence of the Falklands at night was solely in the capable hands of the Phantom F-4 detachment. Routine flying was very varied as there were many military assets available in theatre: the F-4s, C-130 AAR tankers, various Royal Navy ships, air defence radar units, air-to-ground ranges for dropping live and practice ordnance, forward air control (FAC), surface-to-air Rapier units; we even had a practice deployment of Buccaneer aircraft for a week and did some training with them.

Always wishing to make the operation of our aircraft efficient, Mike Beech and I were concerned about the time it was taking to get our formations airborne. Given the very narrow runway and the risk of FOD (foreign object damage to the engine), we could only get one aircraft airborne at a time. Consequently, on one of our earlier flights out of Stanley, Mike and I decided that we would try a close formation take-off to see how feasible it was on the narrow runway. We managed it successfully (at least neither of us crashed) but afterwards we agreed that we wouldn't be doing that again.

During our detachment I was tasked with planning a retaliatory assault for the Harriers against Argentinian airfields on the mainland in the event of an attack on the

Falklands. Since we did not have any reliable fuel flow figures for the aircraft in certain war load configurations, Lez Evans and I carried out a trial to acquire this information. Mike Beech (now a fellow flight commander) and I then produced a plan which was theoretically feasible but was incredibly tight on fuel, using the helicopter landing pads at Kelly's Garden in Falkland Sound as a diversion in the event of not having the fuel to reach Stanley on returning from the mission. It is probably just as well it was never put to the test.

Walking back to the Harrier Flight one day, I was watching one of my Harriers hovering over the AM2 matting runway[13]. The pilot, Ray Coates, drifted over the edge of the runway, the Harrier's jet efflux got under the matting and lifted the runway off the ground right underneath the aircraft. I thought the matting was going to hit the Harrier. ATC advised him to move pretty damn quickly and he put the aircraft down safely elsewhere. The damage to the runway meant that it was necessary for the Royal Engineers to lift up and relay most of the strip which would take some 48 hours of 'round the clock' work. During this period neither the F-4s nor the AAR C-130 would be able to operate; only the Harrier would be capable of getting airborne, but what capability could it provide against an incoming Argentinian attack? Mike Beech and I got our heads together on this and could see that at any one time we should have about half the runway available for take-off (3,000 ft) which meant that with guns and two AIM-9L Sidewinders, we could still lift most if not all of our fuel. The next problem was: what if the crosswind was outside limits for the east/west runway? A further calculation showed that using the taxiways/airport apron, we could use a rolling vertical take-off; given the normally low air temperature (increased power output) and strong wind on the nose (additional lift) we could still get airborne albeit with a relatively small amount of fuel. In the event, the Harriers held the day and night(!) QRA commitment for the Falkland Islands for 48 hours but fortunately we were never called upon to put the plans into practice.

I got on well with the C-130 detachment commander, Squadron Leader Derek Oldham, and we did quite a lot of work with him and his team. Besides air-to-air refuelling[14] we also gave them practice against an air threat and they became very adept at using all their crew members in spotting us and getting the pilots to manoeuvre aggressively to negate a simulated missile launch. On one of my flights with Derek, he put me in the right-hand seat and got me to do the landing. Given the gentle breeze straight down the runway, this went remarkably well but set me up for a fall when I returned to the Falklands the following year.

One aspect of living on the airfield at Stanley which I remember well was the number of rats sharing the accommodation with us. As flight commander, I had my own 12x12 tent, with an inner liner, an iron bedstead with mattress, a table and chair and a very effective arctic heater which ran on aviation fuel. Getting into bed at night was reasonably comfortable but as soon as you turned the light out to go to sleep, you would hear the rats scurrying around between the inner and outer tent. On one occasion they even managed to help themselves to a fruit cake which had been sent from home. We also had a resident rat who shared our aircrew coffee bar that was, inevitably, named Roland. He would happily help himself to bread, toast, butter or whatever was left out and was not at all fazed when we came in, just finishing what he was eating before loping off. It was amazing we all stayed healthy.

On 22 March, the weather was calm and fine although there was a suggestion of fog in the vicinity. The Met man was not happy with the situation, fearing that the fog might move onto the airfield which, since we did not have a diversion option, could present us with a major problem; consequently OC Operations, quite understandably, placed a ban on all flying. After a few hours of drinking coffee and admiring the beautiful sunny skies above the airfield I set out to persuade OC Operations Wing that we should launch a couple of Harriers and, if the fog was as widespread as the Met man thought, we would land straightaway; with some reluctance he eventually agreed. A short while later I took off with Mark 'Leaks' Leakey as my No. 2 and we quickly saw that the Met man was right, the fog was widespread and there was a real risk of it drifting onto the airfield. We dumped fuel to get down to landing weight and just after touching down on the runway I heard Mark call on the radio that his engine had failed. I looked across to see his aircraft at about 800 feet at the end of the downwind leg of the visual circuit with the nose coming up as he tried to maintain height but I could see a high rate of descent starting to develop. Without any delay I pressed my radio transmit button and told him to eject. He left the aircraft immediately and I could see a normal parachute deployment as his Harrier continued descending rapidly out of sight and into the sea just north of the airfield. Pretty straightforward, I thought, another Harrier engine failure and a copybook ejection. What I didn't know was that

The 'Turdis' on the airfield apron at RAF Stanley, Falkland Islands, March 1983.

with no wind, Mark's parachute had descended directly on top of him and he finished up so caught up in the parachute rigging lines that he was unable to climb into his dinghy. He finished up half in his dinghy and totally exhausted. To compound the situation, no-one had told the search and rescue helicopter detachment that flying had started and they had taken the opportunity, whilst there was no flying taking place, to carry out some work on their aircraft and they were unable to get airborne until it had been put back together again. Eventually, after what must have seemed like an interminable time for Leaks, the helicopter got into the air and they cut him out of his entanglement and took him off to hospital for a check-over which revealed nothing untoward[15]. It was interesting to see the reaction to Mark's ejection. The flight-line mechanic who was supposed to be marshalling me in was staring into the empty sky where Mark's aircraft used to be. Having parked my Harrier without his assistance, as I climbed out of the aircraft he said to me:

"Sir, Mr Leakey's just ejected." I replied: "Yes I told him to!"

When I got back to our operations cabin, my deputy was rushing around saying: "We've got to impound all the documents for the Board of Inquiry." But I said, "No we haven't, the Board won't be here for at least three days so we've got plenty of time to make sure everything is in order." I was quite content that there wouldn't be a problem.

During this three-month detachment we had a combined services entertainment show visit the Falkland Islands. One of my pilots, Chris Benn, suggested that we should invite the cast for 'afternoon tea' – the fact that the cast might include some attractive young ladies might have had some bearing on his proposal. It was decided that the 'afternoon tea' would include fresh mussels from the river bed some way out of town. Having commandeered Major Mike O'Flaherty's (our ground liaison officer) LWB Land Rover, we set off to collect the mussels with me driving, Chris navigating and a couple of junior pilots in the back. Having navigated some distance off-road and past a couple of minefields we eventually arrived at the river. At this point, Chris nonchalantly advised me that we now had to drive down the river, assuring me it was only a few hundred yards and, since the tide was out, the water wasn't deep at all. Only slightly concerned that I was endangering one of the best Land Rovers at RAF Stanley, we set off bumping down the river bed unable to see what we were driving over. All went well for the first hundred yards or so at which point the engine coughed and died and refused to restart. Water on the electrics seemed the most likely cause so, gathering any tissues we had between us, I climbed out, opened the bonnet and wiped off the spark plug leads, hastened on by thoughts of rising tides, a written-off Land Rover and a potential court martial. Of course, this was all most entertaining for the junior officers, who assisted by taking photos of my predicament, as any blame would naturally fall to me and not them. Fortunately the engine did decide to come back to life, we did get back and the *moules marinière* were worth the effort.

After three months in the Falklands, we arranged an end-of-detachment all-ranks Harrier dinner night to include not only our 3(F) Squadron detachment but also anyone else at Stanley who had a Harrier background including the station commander, Group Captain Pat King. Pat was a lovely, avuncular senior officer and I had got to know him a little whilst I had been instructing at Wittering and he had passed through the OCU prior to taking command of the station. (I recall discussing whether I should stay in the RAF or not with him one morning, sitting in my car outside the OCU in the pouring rain during this period.) Pat had arrived at Stanley with a firm view that the Falklands was an operational theatre and 'relaxed' behaviour would not be tolerated. I was concerned that our dinner night, at the end of our three months away from home, might run the risk of getting a little too relaxed for his taste. I briefed all the detachment that they were to be on best behaviour at dinner and in our bar tent after dinner until the station commander had left. We had a great dinner courtesy of our excellent RAF mobile caterers with speeches and everyone behaved impeccably; we then moved to the bar for a beer or two. Much to my surprise, Pat King got stuck in and showed no signs of going to his bed so, after a while, everyone started to relax and we had a great night in the bar. Eventually, Pat said he was on his way and I escorted a 'very relaxed' station commander out to his Land Rover. He thanked us for an excellent evening and drove himself very steadily off into the night.

At the end of the detachment, after a flight to Ascension on the air bridge and a VC10 to the UK, Mike Beech and I found ourselves at RAF Brize Norton awaiting an onward flight to Germany. In normal military fashion, this was not going to happen for a few days but we bumped into the crew of a Heavy Lift Belfast in the bar who were going to Gütersloh the next morning to collect a replacement Harrier for Stanley so we begged a lift off them and were unexpectedly back home early the next morning. After the privations of three months living at airport camp it was a delight to be back home once again.

As I moved into my last year on 3(F) Squadron, as the senior flight commander I took over the role of executive officer (deputy squadron commander). It was a good time to take on this role as John Thompson (JT) took over as boss[16] at the same time and largely gave me free rein in organising the day-to-day and week-to-week operation of the squadron – loads of authority without the ultimate responsibility. I relished the role and did a lot of flying in this last year in Germany. Somewhat memorable in this period was JT deciding that he wouldn't go on the squadron exchange with the Portuguese air force near Lisbon in November and sending me off in charge. The weather was truly wet, we didn't do a lot of flying but, as with most squadron exchanges, it was 'socially demanding'; nor did the flight back to Gütersloh go quite as planned.

TRANSIT BORDEAUX TO GÜTERSLOH 1983

As executive officer on No. 3(F) Squadron I finished up leading a squadron exchange to Montijo in Portugal; the boss, probably very wisely, had decided that such an undertaking should be delegated. After nine days of great hospitality from the hosting squadron, sampling the delights of Lisbon, discovering that it rains a lot in Portugal in November and flying very little, our somewhat jaded detachment was ready for the transit back to Gütersloh. The return was planned for Thursday which would give us one quiet night back at base before a Harrier Force reunion kicked off on Friday night for the whole weekend.

The first leg from Montijo to Bordeaux was uneventful but when we landed to refuel, I found that two of the Harriers had unserviceabilities. We contacted the C-130 carrying our ground crew and spares, that was planned to overfly Bordeaux whilst we were on the ground, and told them to divert in to fix the aircraft. Since this was going to take some time, I had to decide which pilots were going to stay with the unserviceable Harriers: I said that I would stay and one of the junior pilots, Boggo, volunteered to stay with me. Eventually, long after the rest of the Harriers had departed for Gütersloh, the engineers declared our two aircraft serviceable, made a mad scramble back into the C-130 and left in haste before the Hercules crew ran out of duty hours. Anyone would think the ground crew were exhausted after nine days in Lisbon and were desperate to get back to their own beds. It was now too late for Boggo and myself to continue back home so we wrapped up the aircraft and were about to try and find some beds for the night when, most unexpectedly, we saw two RAF officers approaching us. It turned out they were on secondment with the French air force with responsibility for the UK supply line of parts for our RAF Jaguars. A couple of quick phone calls from these fine chaps and we had beds in the officers' mess for the night and they had 'pink chits' from their wives to take us out for dinner in town. After the tremendous hosting we had enjoyed from the Portuguese air force, we needed another night out like a third eyebrow. Nevertheless, Boggo and I were not going to let the side down and we had a great meal with our new found friends and eventually finished up in one of their houses drinking some excellent brandy late into the night.

I can't say I felt my sharpest first thing next morning but a quick peep revealed that Bordeaux was covered in a blanket of heavy fog which did not look as though it was going to lift in a hurry. A blessing in one way but also a major concern since we might not make it back to Gütersloh for the reunion that was starting that evening. The French station commander was at breakfast and even offered to get an English breakfast cooked for us; what a wonderful gesture. For some reason we declined. A visit to the Met Office after breakfast confirmed our suspicions that the late autumn fog was very widespread across the whole of mainland Europe.

After a quiet morning with visits to the Met Office every hour, the fog started to disperse and the Met man confirmed that the fog across Europe was now lifting and it was going to be a fine afternoon. Once we had legal diversions, we filed flight plans and made our way out to the aircraft. All was well until we tried speaking to each other on the radio before starting engines when we found that Boggo's main radio was dead although his standby radio was OK. We agreed that we would do everything by hand signals whilst I would do all the radio for both aircraft on air traffic frequencies and Boggo would listen out on his standby radio in case I needed to speak to him. And that is how we got out of Bordeaux although, little did we know it, further problems awaited us en route.

After some 20 minutes in the cruise, I got a call from Boggo on the standby radio to say that he thought his main radio was now working. I gave him the frequency and sure enough all was now well with his radio. However, as we transferred from French air traffic and checked in with Clutch Radar (the military controller in Germany) they asked us to confirm our destination. On telling them it was Gütersloh, he told us that Gütersloh was still red (weather out of limits for landing) and where would we like to go? Despite the forecast and the assurances of the Met man in Bordeaux, the fog had failed to clear from northern Germany. The RAF Germany Clutch airfields near the Dutch border were all in the clear so we elected to land at Laarbruch. By this time, I had a serious problem with my aircraft as the control column would no longer move fore and aft. I still had aileron control so I could control the aircraft in roll but I had to use the electric trimmer to motor the whole tailplane to control the aircraft in pitch. If I could not resolve this, a safe landing was going to be in serious doubt. Sending Boggo down to land at Laarbruch, I descended to 5,000 ft and whilst flying around for some minutes trying to resolve the problem, the controls came free and I once again had full control of the aircraft. Most fortuitously, as we arrived at Laarbruch, OC II(AC) Squadron, Wing Commander Frank Hoare (an ex-Harrier pilot) was just about to set off in his car for the reunion at Gütersloh and gave us a lift back to base. So we did make the reunion after all.

Once a team of engineers came down to investigate over the weekend, they found that a water drain hole in the fuselage had been blocked on my aircraft. The huge amount of rain that had fallen in Portugal had led to a build-up of water in the control runs which had frozen during the extended transit across Europe at high level in very cold temperatures. This had thawed as I had descended to lower levels at Laarbruch, freeing the controls in pitch, fortunately before I had to attempt a landing or ran out of fuel.

Early the next year, the squadron took part in Exercise Mallet Blow in the UK.

EXERCISE MALLET BLOW – 25 JANUARY 1984

The Falklands War was still impacting the RAF long after hostilities ended in the summer of 1982. All Harrier squadrons had been reinforced by pulling qualified pilots back from the training units which had led to some over-manning since, thankfully, unlike the Fleet Air Arm, the RAF had not lost any Harrier pilots during the conflict. The Harrier Force was over-stretched with aircrew and ground crew still detached to RAF Stanley. Nevertheless, we were still required to run a full annual training programme which meant that on 1 February 1984 we would deploy to Decimomannu in Sardinia for three weeks for our annual weapons practice camp. Because of this we were not well placed to support the annual Mallet Blow exercise in the UK. This was a major national training exercise and we would normally detach six aircraft with supporting ground crew to one of the UK's RAF bases for a week to take part.

Since we were keen to participate in Mallet Blow, and all our pilots were now AAR qualified, we hit on the idea of flying the missions out of, and returning to, Gütersloh. As executive officer, and having been party to this idea, I thought it was only fair that I should take on the lead for the first mission on 25 January. We were allocated a four-ship task with a first run attack (FRA) in Otterburn Range in Northumberland followed by a route south into Yorkshire to go through a fighter engagement zone (FEZ), where we could expect to be engaged by RAF F-4s or Lightnings. To ensure getting four aircraft onto the mission, I elected to take five aircraft to the tanker; if all aircraft got airborne and the first four took fuel off the tanker, the fifth aircraft would then return to Gütersloh leaving the four-ship to progress to the target.

On the day of the mission the weather forecast was good for the UK and I spent the morning planning with the rest of the four-ship. The only pilot available to fly the airborne spare (No. 5) was Mike Beech. He was flying that morning, but would join us for the brief; given his experience and abilities, I had no doubt that Mike would be able to take over anywhere in the section if required. We briefed on time and got all five aircraft airborne in good shape to join with our Vulcan tanker at the German/UK airspace FIR boundary. Fuel planning for the mission was quite critical as the Harrier GR3 never had much fuel available, even with the two 100-gallon drop tanks which we always carried. I also had to allow enough time in the plan for the rendezvous (RV) with the tanker and for four of us to refuel before pressing on to the target. We also had to ensure that the Vulcan dropped us off at the right place. In the refuelling plan, it was necessary to allow for the fact that the Vulcan had only one refuelling basket so all aircraft would be waiting around whilst others refuelled. Aircraft that had refuelled would then be eating into their mission fuel whilst others were still refuelling. In the event, the RV worked well, I got my fuel OK and the others took their turn. All was going as planned, we were looking good for our drop-off point in the North Sea and just No. 4 to refuel.

Unfortunately, at this point, the 'big match' pressure on No. 4 proved too much and he was unable to make contact with the Vulcan's refuelling basket. To give him his due, he was a fairly inexperienced Harrier pilot and he also had very limited experience at air-to-air refuelling. After a number of futile stabs at the basket, and with time to the drop-off rapidly approaching, I told No. 4 to get out of the way and let No. 5 take the fuel immediately. Needless to say, Mike made contact with the basket first time, but the refuel rate of the Harrier was always slow, and we were now hurting to make the TOT. Having given Mike as much time as I could to get his fuel, I told the tanker we were departing. As I turned away and started a rapid descent to low level, I could see Mike still plugged in trying to get as much fuel as possible. I told the tanker to wait for us just off the coast abeam North Yorkshire as we would be pulling out of low level on minimum fuel.

As forecast the weather over England was fine – a heavy overnight frost had not cleared, the countryside was blanketed in a white dusting, and the visibility was excellent. The low-level mission proved to be something of a non-event: the FRA went as planned but there were no fighters in the FEZ. However much we checked our radar warning receivers, and scanned the skies around the four of us flying in a wide card formation – two pairs of aircraft, two miles abreast with the second pair three to four miles behind – no-one was out there looking for us. As expected, since he had not managed to get all his fuel from the tanker, a Bingo fuel call came from Mike which meant that he needed to get back to the tanker. Nos. 3 and 5 then pulled up and headed back to the Vulcan. I continued with my No. 2 in the forlorn hope that some fighters might appear but a few minutes later we were on our minimum fuel and pulled out of low level. Switching to a fighter control frequency we requested vectors to our Vulcan.

It came as something of a surprise to find that our tanker was well over 100 miles away, nowhere near where he was supposed to be, and that No. 3 and 5 were only just joining him for fuel now. (It transpired that, having seen the difficulty that No. 4 had had refuelling, the tanker captain had taken it upon himself to escort No. 4 back towards Germany and the FIR boundary and give him some tanking practice, totalling ignoring my request for him to remain just off North Yorkshire!) A check of our fuel state confirmed that we were now in something of a predicament – heading out into the North Sea to meet the tanker would take us away from any diversion airfields and our fuel required to get back to an airfield was increasing all the time. We told the tanker to head back towards us immediately, reduced to our most efficient cruising speed and prayed for a good RV that would not need us to increase speed too much and waste our very limited fuel. By the time we got behind the tanker I had stopped looking at my fuel gauges and I was acutely aware that my No. 2 would have very little fuel as well. Fortunately, I made contact with the basket first time, took on about 500 lbs of fuel then let No. 2 in to fill up. I then hooked up again, filled up, and we then left the Vulcan to make our way back home.

The low winter sun was already setting in the west as we headed south for our Cold War hardened aircraft shelters in north Germany, 70 miles from the Inner German Border. Today our target had been in Northumberland but next week, or next year, it could be in East Germany.

We had a visit by Prime Minister Margaret Thatcher to the Eberhard field site in the Sennelager Training Area during my last year on the squadron. Less than two years after the Harriers' important role in the Falklands campaign, the prime minister appeared genuinely pleased to visit the RAF Harrier Force in its more normal environment. Not least, by this time, most people on the squadron had now done at least one detachment to the islands with some of our pilots having actually taken part in the hostilities.

On one of my last field exercises, as the executive officer I was commanding the squadron's No. 3 field site. On the Friday evening, we ventured off site as normal for a shower and a beer and bumped into the officers from No. 1 site who invited us back to their site for a beer as there was something we had to come and see. Never ones to turn down the offer of a free drink, we were enjoying a couple of beers in their mess tent when the site commander asked us to come outside. Their domestic accommodation was in a forest of sparse but very tall pine trees and by this time it was very dark. In an open area outside the mess tent a young pilot officer, who was attached to the squadron awaiting his next stage of training, was floodlit with torches; he was wearing a flying suit and strapped into a parachute harness attached to a rope. Once we were all assembled the site regiment officer bellowed an order and the poor pilot officer suddenly shot up vertically into the trees to a height of about 40 to 50 feet. In their spare time, the site regiment officer and others had rigged up a system of ropes and pulleys so that on the given order, a team of runners set off with the other end of the rope which pulled the unfortunate victim up into the trees. Now that he was up there, the rope was secured, we all retired to the mess tent for another beer and the pilot officer was left hanging in the tree tops. I have no doubt that today this would be viewed as bullying but it was all done in a good spirit, it was truly hilarious and, I believe, the pilot officer was a volunteer and actually enjoyed the prestige of being the 'test pilot'.

My last field exercise was in June 1984 when my site was tasked with a tactical deployment under cover of darkness with convoys rolling out of base in the late evening, the first time this had been done. All went well and shortly before dawn we were ready to take aircraft into the site. I ordered work to stop so everyone could get some sleep before the first aircraft arrived about 0900. The only thing I had left to do was put my site commander's Land Rover under the cover of a large tree and cover it with camouflage netting. Unfortunately, without the aid of headlights, I failed to notice that there was a deep hole beneath the tree I had chosen and one of the front wheels duly dropped into it and I came to an abrupt stop with the Land Rover canted over at about 30 degrees. By the time I had climbed out of the top side of the vehicle, the NCO in charge of the Royal Engineer detachment on site, Corporal Jones, had arrived telling me not to worry, his boys would have it sorted straightaway, which they duly did. The next day I went to find Corporal Jones to thank him for his prompt assistance and to tell him that I would like to express my appreciation by taking him for a flight in a Harrier during the field exercise. He was almost speechless, his only response being "F***ing hell, sir!" Come the day, Corporal Jones was medically checked for flying, kitted out and brought out to the aircraft with his Engineer Troop in tow. Once strapped into the back seat, photographs taken by his men, and on the intercom all I could get out of him was "F***ing hell, sir!" As we lined up for take-off, I had to ask him to stop repeating "F***ing hell, sir" so I could concentrate on getting us airborne safely, and so it went on for 30 minutes until we got back on the ground again.

One evening in the first week of the exercise, we met a couple of British guys in a bar who were on a summer job operating an ice-cream van in the local area. They asked what we were doing and, having told them, we half-jokingly suggested they should visit our field site where they would be able to sell lots of ice creams. With a cavalier disregard for the security of our flying operation (all the locals knew where we were anyway), we told the guys where our site was and headed back to our tents and thought little more about it. The following week our site, along with the rest of the Harrier Force, was on a three-day TACEVAL being examined by a team from NATO. As the site commander I was out and about on the site keeping an eye on things and answering questions for the visiting NATO assessors, most of whom had probably never seen such a flying operation in their lives. I was suddenly called to the phone to speak with my RAF Regiment officer who was responsible for site security. He informed me that an ice-cream wagon was at the point of entry for the site requesting access and saying that they had been told to come whenever they could. Were they to be allowed to enter? I took one look round at our guys sweltering in their nuclear biological chemical protective suits in the hot June sun, the NATO evaluators with their clipboards, threw into the mix that I was just about to leave the squadron anyway, and told him to let them in immediately. It was a most bizarre sight – an ice-cream wagon parked in the middle of a tactical Harrier field exercise on a TACEVAL. I sent a message to all personnel on site to say they could come along for an ice cream if they were not on essential duty and joined the queue myself. There were a few strange looks from the NATO team but it wasn't long before they downed clipboards and joined the queue as well.

I remember one flight particularly well during my last year in Germany. The AOC-in-C, Air Chief Marshal Sir Paddy Hine, came up from HQ at Rheindahlen with a visiting French 4-star, General Forget. The boss was going to fly the general and, since Paddy Hine had decided he was not going to fly, I would fly No. 2 to JT with the AOC-in-C's PSO, Wing Commander Ian Stewart, a Harrier pilot, in the back of my T4. If anything went wrong with the boss's aircraft, I was to take the general. On take-off JT's aircraft suffered a hydraulics' failure so I had to dump fuel down to landing weight, get on the ground without delay, get the aircraft refuelled and the general in the back of my T4 and get airborne asap. We got airborne in good shape and set off for the North German Plain to carry out a low-level reconnaissance task before flying through Nordhorn Range to drop a practice bomb on a first run attack. As we cleared the range, I was explaining to the general that our pre-planned fighters (RAF Germany F-4s out of Wildenrath) who were going to oppose us in the low-level area wouldn't be there (they would be out of fuel having waited for us and would have returned to base) when I noticed that there were definite signs of air-defence fighter radar activity on the RWR (radar warning receiver). Following our planned route, I soon visually acquired some F-4s operating at low level and realised there were at least five to six of them out there! One versus six was not good odds but given the small size of the Harrier, its excellent camouflage scheme against the North German countryside and keeping low, I had a reasonable chance of getting through. Accelerating to 480 kts, keeping very low, minimising my manoeuvre as 'wing waggle' tended to catch the eye of other aircraft, I was halfway through the fighter cordon when I passed close to a Jaguar. It was then apparent that I had 'bootlegged' into some F-4s that had been booked by the Jaguars – nevertheless, good training for the F-4s to have an unknown threat amongst them. Getting clear to the west I then had to turn back and fly through the fighters again to get back to Gütersloh. During all this I managed to get two good Sidewinder AIM-9L shots at F-4s which appeared to meet with approval from the general in the back seat.

Back at Gütersloh, I had just enough fuel left to let General Forget try his hand at hovering before we landed on minimum fuel and taxied in to be met by the AOC-in-C and the boss. Taxiing back in I had been overtaken by some concern. Bootlegging into the Jaguars' booked F-4s was not a big deal but I did have some worries over the height we had been flying at some of the time to escape detection by the F-4s. The German government had recently expressed concern over the height at which some military aircraft had been operating and had ordered some Super Fledermaus radars deployed to try and identify the culprits. In the German low-level areas we were cleared to fly down to 250 ft, however, in the post-Falkland era all combat-ready Harrier pilots were trained down to 100 ft. I had spent most of this tour qualified at 100 ft and by this time I was very comfortable (perhaps too comfortable) operating at this height. I was now concerned that I might have been picked up by the German radar flying well below 250 ft; the fact that this would have been with a high-ranking French general on board would have made it even more embarrassing. As soon as we climbed out of the aircraft, General Forget put an arm around my shoulders, dragged me over to Paddy Hine and started to tell him what an outstanding fighter pilot I was, with me inwardly cringing about bootlegging the Jaguars' aggressors and mindful of my 'bending' the low-flying height regulations. In the event I heard nothing more about it although, not for the first time or the last, I was left with the uneasy feeling of having overstepped the mark. In this, I don't expect I was very different from many other military fast-jet aircrew.

That summer my posting came through. The boss called me in to his office to tell me that I would be leaving the squadron in August. Even though I hadn't done the basic staff course (a pre-requisite for selection for Staff College), the case put forward that I hadn't had the opportunity to go to basic staff course due to 'operational requirements' (Falklands War etc.) had been accepted. Consequently, I had been selected for the RAF Bracknell staff course the following March with the basic staff course to be completed prior in January/February. Meanwhile I would be taking over the Belize Detachment from August until December. This was all great news and I said I was particularly pleased to be going back to Belize after all these years. John said, "Did I say Belize? I'm sorry, I should have said Stanley!" And so it was back to the South Atlantic once again as Harrier Flight commander – this time for four months.

My last few weeks on the squadron were, as usual, very busy and it was going to prove very difficult to get back to the UK to the planned celebration for my grandmother's 90th birthday before I headed off to the South Atlantic once again. Consequently, tongue in cheek I asked the boss if I could carry out a ranger (navigational training flight) back to the UK for the celebration, which was the official way of asking if I could borrow a Harrier for the weekend. Somewhat to my surprise this was agreed at higher level. When I attended the family party on the Saturday, my Granny Mabel was most impressed that I had 'borrowed' an aircraft just to come to her party that weekend and asked if she could pay for the fuel. She was, however, a little confused over my flying a 'jump jet' and told her friends that I had borrowed a jumbo jet for the weekend!

No luxury cruise liner to Stanley this time, just the joys of the air bridge. As we boarded the C-130 at Ascension for the ten–11-hour flight to Stanley we were all issued with two meal boxes for the journey. There is no hot food for passengers on an RAF Hercules and certainly no alcohol. Whilst meals on long flights are normally the only highlight, we would have to make do with meal boxes. I opened the first one to discover what cold delights I had in store: sandwich, crisps, tin of pâté, chocolate bar and a few other rather uninspiring items. Checking the second box, I found it was exactly the same even down to the same sandwich filling – so absolutely nothing to look forward to. Adding in the noise, vibration and cold of flying in the back of a C-130, landing at Stanley couldn't come soon enough. Still it was good to be met off the aircraft by the Harrier pilots from 1435 Flight who whisked me away and got me settled back into the routine although, by this time the work and accommodation facilities had improved somewhat from my previous detachment in 1983.

In late 1984 there were still a large number of military assets in the Falklands so the flying remained interesting. The Harrier landing pad and strip at Goose Green had not been used for some time and the air commander in theatre was asking questions about its viability. So early on in my detachment I decided that I would take a Harrier in there. All went well for the vertical landing but an inspection of the take-off strip confirmed that it was floating on water and not fit for use. Having allowed for this, I had the correct fuel load to permit a vertical take-off (VTO) from the landing pad. However, as I applied full power to lift off, the aircraft performed a 'roll on VTO', the only one I ever experienced in over 2,500 hours on the aircraft. This was a rare but known characteristic of the Harrier where the aircraft would roll uncontrollably during the lift-off for a vertical take-off. As soon as you realised that full aileron would not stop the roll, the only option was to slam the throttle closed, accepting that the aircraft would be damaged as it was just lifting off and was starting to generate sideways motion. In the event, the outrigger was damaged and this entailed an unscheduled night stop with the resident infantry company at Goose Green. Some of the ground crew were flown in by helicopter and the aircraft was repaired overnight. Next day, using the vertical landing pad and the initial part of the take-off strip which was just about usable, I managed a rolling vertical take-off without incident and returned to Stanley.

Life at RAF Stanley away from work was somewhat limited but I was fortunate enough to meet Rob and Lorraine McGill, sheep farmers who owned Carcass Island in the north-west of the islands. They very kindly invited me to visit their island and I enjoyed a wonderful weekend with them and their two children. What made it very special, besides the outstanding wildlife including elephant seals coming ashore for breeding, was the fact that we were the only people on the island – and normally it was just the four of them. The peace and sense of isolation, especially after life at airport camp, was very special.

SOUTH ATLANTIC SPRING

A memory of Carcass Island

(For Rob and Lorraine McGill)

Spring dawns
And on a cold September day we stroll
On stranded beds of kelp, high washed,
That once were silken carpets laid
On Passages
Through far-flung western isles.

Past elephant cows and bulls we tread
With wonder in the camera's eye
Whilst on the black and fractured foreshore
Falls the rising springtime rip.

Night herons coolly sit and spy
As through the tussock heads we pass;
Out on the lonely open acres
Geese now lay their early eggs.

So back to Camp,
Climbing surely, pausing on the Ovens;
Bitten deep by springtime gales
Which fleeing fast from rocky shore
Now bid us haste

Our homeward way.

The Falkland Islands,
October 1984

One other diversion from the routine was walking in the hills to the west of Port Stanley where the latter stages of the battle for the Falklands took place. It was necessary to be circumspect when doing this because there was still a lot of ammunition and some unexploded ordnance on the ground from the recent hostilities.

Through the McGills I got to know the residents of Westpoint Island and spent a very enjoyable weekend with them as well. Invited back for another weekend with one of my pilots, Mark Hare, the helicopter was unable to get us to Westpoint Island due to fog and, since our seats were required for others returning to Stanley, we asked to be dropped off at Saunders Island where we knew there was an R&R Centre. Arriving there totally unannounced we found the R&R Centre was shut so we were now stuck for 24 hours with nowhere to stay. At the time there was a Falkland Islands Force Standing Order that no military personnel were to leave airport camp at Stanley without sleeping bag, mess tins etc. Since we knew we were headed for comfortable beds and some excellent hospitality at Westpoint, we had somewhat stupidly ignored this perfectly reasonable edict. Standing outside the R&R Centre with no-one in sight and trying to decide what to do next, a young lady suddenly appeared and asked us what we were up to. Having explained our predicament, she said we had better come and have a cup of tea at her house and we could sort out what to do when her husband got back from rounding up sheep. A couple of hours later, her husband appeared absolutely knackered after two days in the saddle rounding up the sheep on the island for shearing. It was quickly decided that we would stay with them, their young son was moved into their bedroom and Mark and I had his room. After an evening meal of mutton and chips, the table was cleared, a cupboard opened and a huge array of spirits produced which was the start of a very long evening of drinking. About one o'clock in the morning, Mark and I were telling tales of our experiences in Belize. Inevitably rum and coke got a mention but there was no rum to be found to accompany the story. At which point our young lady host got up rather unsteadily, left the room and we then heard an engine starting and a vehicle heading off into the night. Asking where she was going, her husband said she was off to the village store to get some rum (all the villagers had their own key for the store). Sure enough, a short while later, she reappeared with the rum and the tales continued. We finished up eating fried penguin eggs to satisfy the munchies before collapsing into bed. It was a rather jaded pair of Harrier pilots that surfaced in the morning and caught the helicopter back to airport camp later that day.

This time the C-130 detachment commander was an ex-Harrier pilot, Squadron Leader Al Holman. He had been an instructor on the OCU at Wittering when I had been a student and it was good to see him again. Over a beer one evening I was waxing lyrical about my landing the C-130 with Derek Oldham the previous year and telling Al how easy it was to land 'Fat Albert'. Of course, Al then said I should come flying with him and show him how it was done. Come the day, once we had fought off my Harriers, Al threw the co-pilot out of his seat and put me in there for the landing. As we approached RAF Stanley, we discovered that the crosswind was right on the limit for the aircraft. Despite my protestations that my landing in these conditions was not a great idea, Al just moved his seat back from the controls, grinned, folded his arms, reminded me about how easy landing 'Albert' was and told me to get on with it. As we arrived at the threshold, with me completely out of control of the situation, Al calmly said, "I have control," sorted it out, made a great landing and turned off halfway down the short runway. The follow-up to this landing was that the station commander (Group Captain Bob Lightfoot, an ex-Lightning pilot and a very nice guy) saw Al later in the day and remarked on his 'interesting' approach to the runway that morning, to which Al replied, "co-pilot's landing, sir." – he omitted to say who the (totally unqualified) co-pilot was.

With a steady turnover of pilots and engineering staff – since the flight was now being manned on a rolling detachment basis from qualified personnel across the RAF – and a focus on making our own entertainment, the weeks ticked over steadily without too much drama until right at the end of my four-month tour.

Author with MBE at Buckingham Palace in March 1985.

In the middle of a force exercise, two of our Harriers were attacking the airfield at Stanley simulating Argentinian fighter aircraft. On his final run-in on the attack at about 500 knots descending to 100 feet, one of my pilots, Ian Wilkes, suffered a major bird strike to the windscreen. Unable to see out of the aircraft and with the engine in surge due to bird remains being ingested, Ian had no option but to eject immediately.

Above: Author with Chipmunk G-APLO (his first solo aircraft) at Perth Scone Airfield, Scotland in August 1969.

Left: Award of Wings by Air Vice-Marshal F. D. Hughes, Commandant of RAF College Cranwell on 24 February 1972.

Below: 511 Squadron Britannia en route to Singapore on the author's 'creamie's benefit'.

Right: Author with Harrier GR3 of 233 OCU, RAF Wittering in May 1976.

Below: No. 1(F) Squadron Harrier GR3s over the Belize Barrier Reef, 1977.

Above: Harrier GR3s and a T4 in arctic camouflage. (Crown Copyright / Ministry of Defence)

Left: Cleaning mussels RAF Stanley, March 1983. Clockwise from front: Mark Leakey, Ray Coates, Major Mike O'Flaherty (army GLO), Mike Allen, author, Gerry Humphreys and Squadron Leader Mike Rudd (visiting Buccaneer Detachment commander).

Below: Prime Minister Margaret Thatcher visiting the Harrier Force at Eberhard field site in September 1983. From the left: four of the squadron's pilots on my Flight (Tim O'Dwyer-Russell, Les Evans, Ray Coates and Murdo McLeod), Mrs Thatcher, Group Captain Dick Johns (station commander RAF Gütersloh and later Chief of the Air Staff), the author, Sir Paddy Hine (C-in-C RAF Germany) and Wing Commander John Thompson (OC 3[F] Squadron and later air marshal).

Above: Author in a Harrier GR5 at RNoAF Bardufoss, November 1991.

Below: The author flying over Burghley House, Stamford in a Harrier GR7.

Left: Princess Diana on one of her two visits to RAF Wittering whilst the author was in command of No. 1(F) Squadron. From the left: Author, Squadron Leader Peter Coyle (SEngO), Princess Diana, Prince Harry and Group Captain Syd Morris (station commander).

Middle: Air Marshal Sir Ken Hayr (ex OC 1[F] Squadron 1970–72) following his final flight in a Harrier prior to retirement in November 1992. At the time, Sir Ken was the most-decorated serving RAF officer with an AFC and bar, CBE, CB, KCB and KBE. Very sadly he was killed flying a de Havilland Vampire at the Biggin Hill Air Show in 2001.

Below: Mixed Harrier formation over the sea. (BAe Systems)

Right: Almost certainly a unique Harrier formation flown over The Wash to mark the change of command at RAF Wittering on 3 December 1992. From the top: GR3 Group Captain Syd Morris (outgoing CO); T4 Group Captain Pete Day (incoming CO); Sea Harrier Lieutenant Commander Jack London (899 Squadron RNAS Yeovilton); GR5 Wing Commander Paul Robinson (OC 233 OCU); Author (OC 1[F] Squadron). (BAe Systems).

Below: Author in a Harrier GR7. (BAe Systems)

Left: Station commander RAF Scampton, September 1994.

Middle: With Carolyn Grace and the Grace Spitfire, RAF Scampton July 1995.

Below: Air Commodore Simon Bostock (commandant CFS) and the author with four aircraft types flown in one day on 31 March 1995.

Right: Freedom of the City of Lincoln.

Below: A Falcon 20 flying in formation with Typhoons.

Whilst this dramatic event was unfolding, most of us on the airfield were in bomb shelters taking cover from the 'Argentinian' air raid. We had left our operations clerk at the operations desk safeguarding the secure documents and the first I knew about the accident was the ops clerk coming in to the shelter to ask me to come and take a call from air traffic. It was the duty aircrew officer who told me that one of our Harriers had crashed in the harbour and though they had seen an ejection they had not seen a deployed parachute.

In the event, Ian had ejected with almost 90 degrees of bank and his parachute had deployed just before he hit the water, very fortunately arresting much of his 500 knots of forward speed. Nevertheless, the impact with the sea resulted in a broken collar bone and a dislocated shoulder meaning that he was unable to inflate his Mae West (life jacket). His personnel survival pack (PSP) was still attached to his backside and this was keeping him afloat, but this meant that he was being held face down in the water. After much struggling to get enough air, Ian was rapidly losing the battle to stay alive.

When Ian ejected he was flying close to the shore where two airmen on a day off had stopped in a boat. They never saw the ejection as all they could see was the underside of the aircraft. However, they started the boat's engine straightaway and motored out to where they could see bits of wreckage in the water; notable amongst this was a yellow box (the PSP) which was bobbing up and down in the water. On reaching this and trying to pull it out of the water, they were very surprised to find a pilot attached to it! The medical staff met Ian at the jetty but he was in such pain that they had to give him two shots of morphine before they could move him to hospital. By the time we got to see him, an hour or so later, he was definitely on another planet. Ian was flown back to the UK a few days later, once his condition had stabilised, and didn't fly again for many months. As the flight commander with responsibility for Harrier operations at RAF Stanley, I was interviewed by the Board of Inquiry at Ascension Island as I travelled back to Germany at the end of my detachment and they flew south to start their investigations. And that was the end of eight-and-a-half years of continuous Harrier flying. Back in the UK at the end of December, I was surprised and delighted to learn that I had been made a Member of the British Empire (MBE) in the New Year Honours List 1985.

CHAPTER 8
STAFF OFFICER, STAFF COLLEGE AND REFRESHER FLYING 1985–91

At the start of 1985 I attended the six-week basic staff course at RAF Bracknell as a precursor to staying at Bracknell to attend the advanced staff course for the rest of the year. In late January, I was called out of a syndicate discussion to take a phone call from the Personnel Management Centre at Barnwood in Gloucester. My desk (posting) officer, Squadron Leader Neil Taylor, told me that I had a posting to join them in Gloucester in March as a desk officer. My immediate thoughts were 'great, that is a really good posting' and then 'why is he telling me this now when I know I am at Bracknell for the whole of this year?' A bit slow on the uptake, I then realised that my posting was for this March (the present incumbent of the post had been moved at short notice to a military assistant appointment) and that I would not be staying at Bracknell for the advanced staff course after all. So in March I joined the staff of the Department of Personnel (Air) at Barnwood in Gloucester.

After a very quick handover, I found myself responsible for the postings of some 450 pilots flying Harrier, Jaguar, Tornado and Buccaneer aircraft plus simulator staff and other officers in ancillary posts. The office was shared with three other squadron leaders who looked after the rest of the RAF's junior pilots; we also had three civil servants working with us as clerical staff. Having taken over the job, and after a somewhat stressful first week on my own, on Friday afternoon the phones stopped ringing and I decided to clear out a filing cabinet I had inherited from my predecessor. Amongst other interesting documents I came across was a boarding for an exchange tour flying the US Marine Corps AV-8A (Harrier). I found that I had been boarded for the appointment but had been discounted as I was not married[17]! By this time, the clerical staff (three young ladies from Gloucester) had completed their work for the week and came to join me. The chat quickly turned to my flying background and it became apparent that the other squadron leaders in the office had been telling the girls that my coming off Harriers was really something, only the best went to Harriers and I was obviously something special etc. etc. Having cottoned on to this, I thought I would shoot a bit of a line and explained that – not to be repeated – the psychological demands of flying Harriers were such that after a few years it got to some pilots and, especially on field operations, it was not unknown for pilots to have a couple of fingers of whisky before going flying in the morning. After much ooh-ing and ahh-ing about this we departed for the weekend on good terms.

I had been told by the other desk officers not to expect the clerical staff to make you coffee but, if you were in their good books, they might oblige. I was therefore very pleasantly surprised on the Monday morning, as I took the first call from one of my new customers, to find a coffee placed in front of me by my clerical assistant. Thinking how my chat with the girls on Friday had obviously made the right impression, in mid-call I picked up the coffee to find it had been very liberally laced with whisky. Turning round, I found the three clerical staff and the other three squadron leaders all enjoying me being hoist on my own petard. No way out, I downed the coffee and by 0915 on Monday morning I was certainly feeling the effects of the alcohol at which point the air commodore's PA came into the office to tell me that I was required now for my arrival interview with the director. In something of an alcoholic haze, I recall the air commodore telling me that within a month he expected me to know all about the officers I was responsible for – some 450 people – quite a tall order. The way I tackled this was to rewrite the whole book I had inherited from my predecessor (pre-computer days) with all the pilots listed by squadron with their initials and background and, in some cases, cryptic notes concerning their promotion prospects. Even to this day, if I hear the names of some of those pilots, I can still recall their initials and the squadrons they served on.

The department was a very interesting place to work and I was fortunate to have a great bunch of colleagues to work with. My main role was to maintain all the squadrons and flight simulators I was responsible for manned with the correct number of pilots, keeping the experience level equal across all squadrons of the same aircraft type and ensuring that key junior officer appointments, such as weapons instructor and flying instructor, were filled. All this had to be done whilst trying to meet the personal wishes of all my customers for the progress of their careers. At this time in the mid-1980s, I was also managing the reduction of the Jaguar Force and the build-up of the Tornado GR1 Force which often necessitated sending Jaguar pilots through Tornado training (a six-month process) to be ready to join a Tornado squadron on time as it formed up in the UK or Germany. I also inherited another problem which was that the RAF was training more Jaguar pilots than it actually needed, especially bearing in mind the rundown of that force. Consequently, we had to 'hold' Jaguar pilots on the operational conversion unit at Lossiemouth until space came available to feed them through to one of the front-line squadrons.

Quite early on in my tour I was called into the group captain's office. He told me that I needed to find two candidates on my desk who met certain criteria: flight lieutenant approaching promotion to squadron leader, offensive support qualified weapons instructor and married. I returned to the group captain's office later with the files for two candidates. Having checked that they were suitable, he told me that he had a meeting the next day with the Chief of the Air Staff and that he was taking the files with him up to London. Two days later I was summoned back to be told that I should arrange for Flight Lieutenant X to leave my desk in six months' time, I was not allowed to ask where he was going and I should not discuss this with anyone else. It was only some time later, that I found out that Flight Lieutenant X was the RAF's first pilot to be involved in the USAF's Stealth programme.

I was also able to influence some issues within the Harrier Force that concerned me. It was not unusual for pilots in Germany to request a re-tour at Gütersloh at the completion of their three years and they were normally given a two-year extension on the same squadron. I felt that this perpetuated the division between the two Harrier squadrons at the base and I thought it would be much better to cross-post officers between squadrons. Consequently, the first time this cropped up, I spoke with the squadron commanders, got their buy-in, and cross-posted two pilots. Come the day of their postings, it was a cold and foggy day (not unusual in mid-winter in northern Germany) with no flying taking place. By arrangement, the pilots were bundled into Land Rovers at their squadrons and driven down to the mid-point of the runway where the exchange of pilots took place – shades of 'Checkpoint Charlie'. In addition, in an attempt to try and stop the Harrier Force becoming too insular, I also arranged for some Jaguar and Tornado pilots to move onto the Harrier.

The department was small enough that everyone knew each other pretty well and dealing with postings and careers generated a shared confidentiality and mutual trust between desk officers. As the only Harrier man working there at the time, one of the wing commander desk officers came into the office one day and put the files of two Harrier wing commanders in front of me and said: "Which one for the OCU and which one for OC Operations?" The answer was obvious to me and I told him what I thought. A few weeks later the postings came out as I had advised.

In the latter part of this tour, my group captain was Frank Hoare – an ex-Harrier and Jaguar pilot and a former leader of the Red Arrows – who was a lovely guy to work for and very supportive. With a pilot retention problem that was beginning to build, a decision had been taken over the employment of pilots who had requested premature voluntary release (PVR). This was simply that if a ground appointment had to be filled (e.g. simulator instructor), if possible, the department would post a pilot who had PVR'd into that appointment and keep pilots who were still committed to fulfilling their terms of service in a flying appointment. Whilst this necessarily raised questions over the 'return of service' cost of training in some cases, there was no doubt within the department that this was the correct policy given the difficult circumstances. One day I was on the phone to a wing commander at one of the tactical weapons units explaining that one of his pilots, who was on a PVR, was being posted to a ground appointment to see out his time before leaving the RAF. Whilst the call was in progress, Frank Hoare walked in and started going round the desks, talking to each of the other desk officers about what was going on at their desks. He was obviously listening in to my side of the conversation I was having just as the wing commander was telling me in no uncertain terms how he disagreed with the department's PVR policy. Frank indicated that I should come into his office once the call was finished which I duly did. He wanted to know what the call was about and when I explained he said he would come back to me about it. About ten minutes later he came into the office to tell me that that particular wing commander would never be in touch with me again direct and that he had been told that all his future personnel dealings would be conducted through Group Headquarters and not direct with ourselves in the MoD. It was great to have top-cover like that.

Much of our work involved customer interface on the units and I would spend quite a lot of time, often with the navigator desk officer, visiting the front-line stations to meet with the squadron commanders and the aircrew. Having an understanding boss, I also got checked out to fly the Chipmunk with the Air Experience Flight at Filton near Bristol which provided me with the opportunity to keep in the air quite regularly. This provided a slightly embarrassing episode when I went down to Filton one afternoon in November and the flight commander, Flight Lieutenant Ken Miles, asked if I could take an aircraft up to RAF Shawbury (in Shropshire) and bring another one back. Since all of our air experience flying was done within 20 miles or so of Filton, the opportunity to fly cross country up to Shawbury and back was not to be missed. A check of the weather and NOTAMs, a line on the map and off I went with Ken reminding me not to get back late as it would be getting dark early. Needless to say, I got delayed at Shawbury. Halfway back it was getting decidedly 'dusky' and I still had another 25 minutes to go. Of course I didn't have a torch with me and I now found I couldn't find any way to turn on the cockpit lights. Suffice to say I got back on the ground safely at Filton even if I could barely read the airspeed indicator on the approach. Poor old Ken was waiting for me looking rather concerned.

During this time, I decided to keep flying in another way – by instructing at one of the flying clubs at Gloucester Staverton Airport. One of the schools was keen to have my assistance and provided me with a few hours flying on the Piper PA-38 Tomahawk to prepare myself to be examined for a civilian instructor's rating. I arranged the test at Cardiff with a venerable CAA examiner called Cliff Hubbard who I had never met before. The week of the test I was flying RAF Air Experience Flight Chipmunks at RAF St Athan, just down the road from Cardiff Airport and on the day, owing to a late start due to the weather, I was still in a Chipmunk when I should have been on the road to meet up with Cliff. By the time I landed I had two options: go back to the officers' mess and get changed before going to Cardiff and arriving late, or driving there now, still in my flying suit, and just make it on time. I thought it would be preferable to arrive on time even if I was still in uniform rather than presenting myself late to the examiner. I arrived at the flying club to meet Cliff just in time; he took one look at me, his face lit up and he said, "Still in the RAF are you? Great, let's go and brief," and off we went into a briefing room. Having not instructed at a basic level of flying since Linton-on-Ouse, I had spent some time preparing for this test but was not exactly brimming with confidence in this civilian environment. Once in the briefing room, Cliff asked me my background. Whilst I was explaining that I was an A2 QFI on Jet Provost and Harriers and an instrument rating examiner, he got out the examination sheet, said: "OK – ground briefing stalling one." By the time I had thought, that's OK, I know that one all right, he was writing 'Excellent', signed the form, turned to me and said, "Right, let's go flying," and off we went.

Cliff pointed out a PA-38 and told me to climb aboard whilst he went off to sign for the aircraft. I hadn't flown a Tomahawk for some weeks at this point so I had a few minutes in the cockpit to remind myself of the checks and engine-starting procedure. I waited until Cliff joined me and then started the pre-start checklist. After a moment he said, "It's OK we haven't really got time for that," and he took over, started the engine, started taxiing and then called the tower for taxi clearance. In the event we went off and spent a very nice hour looking at the stalling and spinning characteristics of the Tomahawk because, as Cliff said, "Civilians don't like that sort of thing," (it turned out he was, of course, ex-RAF). He got me to patter (talk through) a circuit back at Cardiff and we were done. Cliff had to rush off to the garage to get his car before they closed – and I was now a CAA flying instructor.

I left Barnwood at the end of 1986 to attend the Joint Services Defence College (JSDC) course at the Royal Naval College at Greenwich in lieu of the RAF advanced staff course at Bracknell. This was a privileged six months studying with RAF, army, RN and civil service colleagues. Besides the graft that goes with attending any staff course, Greenwich certainly had its lighter moments. On the RAF visit, I elected to head north to Kinloss and Leuchars to learn about the Nimrod and air-defence forces. Boarding a VIP Andover at Northolt to fly to Kinloss at the start of the visit, the weather was very unpleasant and some of my fellow students were looking decidedly anxious about getting airborne into a storm. I asked the corporal steward who the captain was and, having got the name, my non-RAF colleagues (who knew I had been in personnel management) asked me if I knew him (which I didn't):

"Yes," I said, "I'm sure it will be fine."

"What do you mean?" they asked.

"Well, he was treated for alcoholism a couple of years back but since he's back flying, I'm sure he's now been rehabilitated and passed fit by the Medical Branch."

It was all I could do to keep a straight face and, of course, we had a good trip north with silver service afternoon tea even if it was rather turbulent.

On the last week of the course we went on a visit to Berlin. Arriving at our accommodation on a very hot and sultry evening in early July, four of us who had run the London Marathon earlier in the course decided we would go for a run before having dinner. After 30 minutes we had had enough and had certainly worked up a good sweat. We were in a huge park but had no idea where our hotel was. Coming round a corner by a wood, we suddenly arrived at a lake, the side of which was covered in naked bodies sunbathing with more people swimming in the water with nothing on. One look at the cool inviting water was enough for us and we threw all our clothes off and dived into the lake with the rest of the nudists.

In spring 1987 I had been notified of my promotion to wing commander in July and was given a posting notice to HQ 1 Group as Wing Commander Offensive Support. However, at this time, a review of personnel management within the RAF was being undertaken which criticised the short tours being undertaken by desk officers at the Personnel Management Centre (my tour had been just 21 months). Consequently, one of the recommendations was for longer tours to provide more continuity and therefore more expertise in the personal management of the aircrew. As a result, my posting was changed to send me back to the Personnel Management Centre to take over from my wing commander at Barnwood.

In July of 1987, I returned to the department I had left six months earlier to assume responsibility for the posting of all junior officer pilots and all junior air traffic control officers in the RAF, some 2,600 personnel at that time. My staff consisted of six squadron leaders and an administrative support team of four civil servants. This was a fascinating and rewarding job, spanning as it did the totality of the RAF's flying operations. Inter alia, this gave me the opportunity to visit the full range of stations and aircraft types that were in service at this time.

Naturally there were some major issues that needed to be addressed during the three years I spent in this appointment, the first of which was the retention of the RAF's pilots. In the late 1980s there was a huge demand for qualified pilots to join the airlines. The option of leaving the RAF after a couple of tours (typically five to six years after training) and moving to the airlines, with their higher earning potential and family stability, was understandably very attractive to many RAF pilots. As a result, my department was subject to increasing demands for PVR which, even though this was controlled through a quota system which I managed, made manning the front line with the correct number of experienced pilots difficult to say the least. We tackled this through a number of initiatives including: a briefing to the Air Force Board; emphasising to our pilots the potential benefits of staying in the RAF rather than moving to the airlines; and the staffing of a scheme to assist pilots to transfer to the airlines on the completion of their initial commitment of 16 years' service (this would include more recognition of their service flying by the UK Civil Aviation Authority for the issue of a commercial flying licence). Meanwhile, to address the shortfalls we were beginning to see in cockpit manning, I reversed a previous decision made by one of my predecessors and encouraged the retention of fast-jet pilots, after their initial 16 years, by giving them the opportunity to transfer to transport aircraft, helicopters or instructing for their remaining service. I also discovered an Air Secretary's staff instruction, which no-one appeared to be aware of, covering re-employment in lower rank. This enabled a senior officer (who was by then removed from the cockpit by virtue of his rank) to return to full-time flying duties in a lower rank but retaining his full pay (a combination of pay for lower rank and part pension). Whilst the take-up on these schemes was as expected limited, given the cost of training a military pilot, every pilot retained represented a huge saving.

To return to the briefing to the Air Force Board, this had come about because I had raised the issue of pilot retention as being in need of urgent attention. This was agreed within the department (1-star level) and briefed up to the Air Secretary (2-star) who was new in post. He concurred and a briefing date was agreed with the Air Force Board Standing Committee. I was given the role of OHP slide operator – a wing commander was not be entrusted with a speaking part in such august company, even if he was closest to the issue. The time allocated to this item was understandably limited and unfortunately time ran out before it had been fully discussed and any conclusion reached. In quickly summing up, the Chief of the Air Staff turned to the Air Secretary and said, "So what you're saying is that there is a bit of a problem but no action is required at the moment?" Being new in post and a lowly 2-star amongst 3- and 4-stars, the Air Secretary agreed and we left the room having achieved nothing other than flagging up the issue. I was so frustrated by this unsatisfactory outcome that I took the very unusual, and undoubtedly naive, step of writing direct to the CAS. After some weeks I received a somewhat dismissive letter from CAS along the lines of history being full of examples of people rushing in with solutions when delay would have served better. (I also received a token slap on the wrist from my air commodore for writing to CAS direct!) I am still of the view that there was much that the RAF could have done at that time to stem the outflow of pilots and encourage retention e.g. the acceleration of the schemes to assist the transition of pilots to commercial flying based on a full return of 16 years' service.

Halfway through my tour, we came up against a problem of a serious mismatch between the number of pilots coming out of training and the front-line requirement (the excess of Jaguar pilots that I had dealt with during my previous tour in the department was really another manifestation of this issue). The reason for this mismatch was simply due to the changing pilot requirement: with a three to four-year lead time to produce a pilot on the front line, decisions taken on reducing front-line squadrons or removing aircraft from service prematurely in this timeframe would, inevitably, lead to an excess of pilots coming out of training. There was no simple answer to this as defence reviews and political decisions on force reductions tend to happen in a very limited timeframe. However, since something had to be seen to be done, a huge standing committee was set up including representatives from across the recruitment, training and posting areas, to 'ensure that this never happened again'. Consequently, every month for the second half of my tour, I had to travel up to London for the day to attend this committee. Although this forum raised the visibility of issues across the totality of the recruitment and training spectrum, I do not recall the committee making any worthwhile contribution to resolving the fundamental problem. In truth, given the short-term nature of defence reviews and cuts, there really wasn't a solution apart from making pilots redundant which did happen most notably in 2011.

Although my appointment was largely desk bound, I got out of the office when I could to stay in touch with the officers I was responsible for and to get back in the air whenever possible. A couple of flights I remember particularly were one in an Andover and the other in a Wessex. I remember the Andover flight as it was a flight calibration of a TACAN beacon and probably the most boring flight I have ever been on from the flying perspective as a passenger. The irony, although of course I didn't realise it at the time, was that I would be flying the very same calibration flights myself some 15 years later in a King Air. To be honest, it did give me lots of time to talk to the crew and I was certainly hosted well as we went round and round in circle for what seemed like hours and probably was. The Wessex flight was in Northern Ireland and somewhat at the other end of the spectrum. My rotary wing desk officer organised the visit to Aldergrove and I was scheduled to fly on a 'routine' flight calling in at a number of army bases. I met up with the pilot who took me off to get kitted out with a bone dome and a *flak jacket*. As we went out to the aircraft, he asked me if I would be able to get the aircraft back on the ground if anything happened to him whilst we were airborne? Given my five trips on the Whirlwind before starting my Harrier conversion, I assured him that it wouldn't be a problem. All went well and my latent helicopter handling skills were not called upon even though, on the approach into South Armagh, my pilot did advise me that there had been some small-arms fire there the previous week.

In early 1990, the director informed me that I would be leaving Barnwood at the end of the summer for refresher flying and taking command of No. 1 (Fighter) Squadron the following year; however, I was not permitted to tell anyone until it was announced officially. One of the perks of working in the posting branch was that you usually got an early heads-up on your next move. Having completed just over three years in this appointment I left Barnwood in the late summer of 1990 for refresher flying at the Tactical Weapons Unit (TWU) at RAF Brawdy.

The course on 79 Squadron was tailored for qualified pilots and navigators returning to flying duties after a stint in ground duties like myself and for front-line pilots returning to the Hawk to instruct at the TWU. Day one in ground school was interesting as the chief technician who was instructing us on the Hawk's aircraft technical systems was covering them very fast. Whilst there was some similarity with the Harrier (both aircraft coming from BAe Systems), I was finding it difficult to keep up and was asking questions to clarify certain issues. Eventually the chief technician said, "You have flown the aircraft before, haven't you sir?" to which I replied no as my original fast-jet training had all been on the Gnat and the Hunter. He then apologised saying that since the Hawk had now been in service for 15 years, he hadn't come across anyone on the refresher course that had not flown the Hawk previously. Once he slowed down, things got a lot easier and we had covered all the aircraft systems in one-and-a-half days. After a few simulator details I got to fly the Hawk which was a delight. It was a straightforward aircraft with a very clean airframe giving it a good turn of speed, despite a fairly low-powered engine, and excellent fuel economy (unlike the Harrier). The handling was largely viceless and crisp, excellent in close formation, yet sufficiently stable, and relatively easy to fly accurately on instruments.

It was good to be back flying again and I quickly progressed through the refresher course; in fact, I was progressing so quickly that I could see that I was going to complete all the exercises well ahead of my scheduled end-date. The staff were not at all concerned with this and said that, if I was happy, once I finished the course I could stay on as honorary staff and help them out, a highly unusual arrangement. It was good to find that the course was pitched at a level that acknowledged one's previous experience despite, in my case, having been away from full-time flying for well over five years. I recall one flight in Scotland leading a pair with a single Hawk as offensive air against us, when we were engaged at low level by a section of Tornado F3s. Having had a fairly extensive fight with them, we disengaged to return to our route, but unfortunately, I was now 'temporarily uncertain of my position' (aka lost). Pressing on to pick up a landmark, which I did fairly quickly, I then realised that we had just inadvertently entered the Aberdeen Civil Air Traffic Zone at 250 feet without clearance. We executed a sharp turn to get out of there and got on with the rest of the exercise. Once back on the ground, the staff supervisor who was flying in the No. 2 aircraft said he would make our apologies to Aberdeen (he hadn't picked up on our position either). However, being the formation leader, and the senior pilot in the formation, I told him I would take responsibility for my error and would call them. The response from the Aberdeen controller was they had seen us, thanks for calling, and please try not to do it again.

After a month of excellent flying at Brawdy I had completed the course covering a Hawk conversion and instrument rating, air combat, weapons (bombs and guns) and attack profiles with an air threat. With a month to spare before starting my Harrier refresher training and Harrier GR5 conversion, I then spent the next three weeks at Brawdy flying as a staff pilot carrying out forward air controller training and 'flag tows' for air-to-air gunnery (a potentially dangerous occupation which I could well understand the staff being very happy to hand over to a wing commander who was only too happy to be back flying again).

At the end of November I returned to RAF Wittering to refresh on the Harrier GR3. Although I had really enjoyed flying the Hawk, it was only when I got back into the Harrier that I really felt at home again after such an extended period on the ground. My left hand was always trying to find the nozzle lever on the Hawk. After the GR3 refresher, it was back to ground school to learn about the new Harrier which was quite different from its predecessor having been heavily redesigned by McDonnell Douglas in the States. At this time there was no two-seater or simulator (the Harrier T10 was still some years off, as was the state-of-the-art eye-slaved simulator). It was an interesting experience getting airborne in the GR5 for the first time without the benefit of dual training or simulator; one of the OCU QFIs was available on the radio if you had any issues, otherwise you were, literally, on your own. It was one thing doing this as an experienced GR3 pilot like me, but quite different for the first tourists who had to do this following a very brief Harrier GR3 conversion to learn the rudiments of VSTOL flight. I applaud the pilots who did this.

By late February I was well into my Harrier GR5 conversion and returned from my first range detail one Thursday afternoon to be met by Squadron Leader Geoff Glover, OC Basic Squadron on the OCU. Geoff explained that my start date on No. 1(F) Squadron had been put back by a month due to the squadron's deployment to the States and how would I like to do some instructing on the OCU? I reminded him that I had effectively been away from flying for well over five years and was only just current again on the GR3. All to no avail, Geoff told me that he was sure I was up to it and he had a T4 ready so we could go straight out and start a QFI refresher course that afternoon. After two more trips with Geoff the following day, I was signed off as 'competent to instruct' on the T4 and flew an Exercise 1 with one of the new courses on the Friday afternoon. It was a strange experience being back on the OCU as a 'guest' instructor, nine years after leaving there for Germany. During the month I instructed on the OCU, I covered all the basic VSTOL exercises with various members of No. 4 GR5 Course and renewed my instrument rating examiners qualification so that when I joined No. 1(F) Squadron I was, once again, a Harrier QFI/IRE which, in the event, turned out to be very useful.

CHAPTER 9
OFFICER COMMANDING NO. 1 (FIGHTER) SQUADRON 1991–94

Having prised myself away from QFI duties at the OCU, I cadged a ride on an RAF Hercules outbound for Nellis USAF base in Nevada where the squadron was taking part in Exercise Red Flag. This gave me a welcome opportunity to get to know some of the squadron's personnel that I didn't know already and to see them in action before I took over command from Wing Commander Iain Harvey back at Wittering in May. After a short detachment to Denmark in June, the first major and most unexpected event in my tour was the grounding of all Harrier GR5 aircraft on 30 July. This was the result of three aircraft fires in short succession which had been caused by the breakdown of Kapton electrical wiring leading to arcing. Since it was apparent that there was not going to be a quick fix to this issue, with the agreement of the station commander, Group Captain Syd Morris, I borrowed two Harrier GR3s and another T4 from the OCU (to add to the one we had already) so that the squadron could keep flying albeit in a limited fashion. Fortunately one of my flight commanders, Mark Green, was, like me, current as a Harrier T4 QFI so we set about back-converting all the squadron's pilots onto the GR3.

Taking command of No. 1 (Fighter) Squadron from Wing Commander Iain Harvey, RAF Wittering 1991.

The Vintage ACMI in October 1991. Two of our ground crew (Flight Sergeant David Ellis and Chief Technician Dennis Horseman) are holding up a bike with the squadron badge to hide the 233 OCU motif on the Harrier GR3.

A number of the pilots had lots of experience on the older marks so they were not a concern, but the newer pilots, who had joined the Harrier Force straight onto the GR5, only had a few hours on the T4/GR3 leading in to their GR5 training. This was a safety concern in two main areas: the GR3 had a very weak auto-stabilisation system in the VSTOL regime and it had poor flight instruments (both head-up and head-down) compared to the GR5. There were a number of other shortcomings which these inexperienced Harrier pilots would have to contend with such as an unreliable inertial navigation system and a much-reduced fuel load compared to the GR5. In the event, everyone coped admirably and we even managed a detachment to the NATO air combat manoeuvring instrumentation (ACMI) range in Sardinia with our two T4s and two GR3s. Our opposition on the detachment was RAF Phantoms and Jaguars, both of which, like the GR3, were coming towards the end of their service life so the detachment was dubbed the Vintage Aircraft ACMI.

Another early surprise when I took over the squadron, but by no means as far reaching as the grounding of the GR5, came in a visit to my office by Major Mike Howes, my army ground liaison officer. The conversation went something like this:

Mike: "So what do you want to do with the 'secret squirrel' fund, boss?"

Me: "What secret squirrel fund might that be, Mike?"

Mike: "Ah, so Haggis [my predecessor] didn't say anything to you about this?"

Me: "Err, no!"

Mike went on to explain that the previous year, when the squadron had been involved in the Farnborough Air Show, they had acquired quite a lot of sponsorship money from industry and from selling off squadron memorabilia to the public. Instead of lodging the money with RAF Non-Public Accounts at Wittering as required, Iain Harvey had decided to retain control of this money and had placed it in a private bank account in town in Mike's name. This was not exactly a heinous crime but one which would raise eyebrows if it was to be uncovered. I was therefore left in something of a dilemma. Should I pass the buck up to the station commander, which would almost certainly lead to an investigation with questions asked of my predecessor (who was in the process of leaving the RAF) and Mike? Or should I keep it in-house and dispose of the matter quietly myself, accepting that if it came out, I would undoubtedly get caught in the collateral damage?

After due consideration and having checked with Mike that the 'fund' was a closely guarded secret, I devised a plan with Mike to spend the remaining few thousand pounds and close the fund down. The following year, we commissioned Mike Rondot, the well-renowned aviation artist (and ex-RAF fast-jet pilot), to paint a picture of a GR7 getting airborne on a dark and stormy night to commemorate the squadron's conversion to night operations in 1992–94. The excellent painting, entitled 'Night Attack' is now owned by No. 1(F) Squadron. Having paid for the painting, the remaining money was consumed by all ranks of the squadron at the next beer call, without most people knowing where the money had come from.

In September 1991, Squadron Leader Ashley Stevenson had a serious bird strike flying at low level over the Yorkshire Wolds in one of the T4s; Cadet Kate Saunders, a Cambridge undergraduate was in the back seat on an air experience ride. The bird struck the aircraft canopy, striking Ashley on the visor and oxygen mask, ripping the mask from its secure mounting and rearranging his teeth in the process. With bird remains and parts of the shattered canopy entering the engine air intake, causing the engine to go into surge with a total loss of power, and unable to see anything due to bird remains on his visor, Ashley took the only course open to him which was to eject immediately. As he was unable to speak to Kate, she very quickly got the idea as she saw Ashley's ejection seat leave the aircraft right in front of her! Although Kate ejected without delay, the aircraft was by then so low that she landed very close to the aircraft in a field that had been set alight by burning fuel and suffered serious burns as well as multiple injuries. Despite his own debilitating injuries, Ashley quickly came to Kate's aid and was awarded a Queen's Commendation for Brave Conduct for his actions in saving her life. This had been Ashley's second ejection in just 11 months, the first being in a GR5 following a catastrophic engine failure at 22,000 feet.

As expected, the grounding of the Harrier GR5 took some time to resolve. As we were the only UK-based GR5 squadron, BAe Systems were given access to my squadron's aircraft to identify the issue and work out a fix. Having examined one of the aircraft, they quickly identified that the problem lay with the routing of some of the electrical cables where it was subject to flexing of the airframe in flight; they then moved on to the next aircraft to confirm the issue. Unfortunately, this showed that the cable routing on this aircraft was different from the first one. It subsequently transpired that there had been a breakdown in the build quality of these aircraft and that no two aircraft had the same cable routing. A complete rework of some of the wiring looms was required to bring all the aircraft up to a common standard. As the aircraft were slowly released back to the squadron they had a limitation of 'day operations only' due to concern over the integrity of the electrical system – a great confidence boost for those of us who were going to fly them! Once we had a few aircraft available, we were then able to meet a planned detachment late in the year to Bardufoss in northern Norway, one of our NATO reinforcement bases.

Pre-positioning at Kinloss the afternoon before, we flew on to the Norwegian air force base at Ørland the next morning for fuel before continuing to Bardufoss. Airborne out of Ørland, my No. 2 found he could not retract his undercarriage. Giving his aircraft a visual inspection, I quickly identified that he had left the nosewheel door

Pre-flight external checks on Harrier GR5 at RNoAF Bardufoss, November 1991.

accumulator T-handle out[18], as a result the undercarriage would not retract. Flying low level up the fjord and out of sight of the control tower, we dumped a lot of our fuel to get the aircraft down to landing weight, returned to the airfield and landed. After a quick turn round, a refuel and, in the case of my No. 2, remembering to reset the nosewheel T-handle, we were airborne again and soon up to height and en route. As we levelled off, the sun was setting rapidly; flying north it was evident that we were never going to make it to Bardufoss before dark, an issue I was aware of but had pushed to the back of my mind during the last couple of hours of somewhat hectic activity. So the options were: return to Ørland, land in daylight and spend the night and possibly the weekend there; or, press on to Bardufoss and arrive in the dark (in contravention of the current 'day only' restriction on the Harrier GR5), meet up with the other pilots and enjoy our planned weekend in Tromsø (the 'Paris of the Arctic Circle'). Needless to say we pressed on. By the time we arrived at Bardufoss it was fully dark and we were faced with an instrument approach on a very steep glidepath, descending into a valley with high mountains on both sides. Suffice to say we both got down safely, even if my approach was somewhat ragged after a rather busy and fraught day. I was very relieved to relax over a few beers that evening and we all enjoyed an excellent weekend in Tromsø.

In September of my first year back with the squadron, around the time of the Battle of Britain weekend, I was sitting in the crew room having lunch with some of my pilots when a State 1 (a crash or imminent crash) was declared by air traffic. The

Harrier GR5 AIM-9G Sidewinder firing Aberporth Range. (BAe Systems)

No. 1(F) Squadron pilots and senior engineering officer at RAF Valley for a missile practice camp in January 1992.

tannoy message stated that a Hurricane aircraft was inbound with engine failure. We rushed out of the crew room to see the RAF Battle of Britain Memorial Flight Hurricane approaching the airfield from the north over Stamford; it was slowly losing height and intermittent clouds of smoke were coming from the exhausts. As the aircraft turned to line up with the runway at a very low height, the lower wing stalled, the wing dropped further and the Hurricane cartwheeled before coming to rest. From where we were standing, we could not tell whether the aircraft had landed the right way up or upside down due to the runway sloping away from us. Seconds later the aircraft went up in flames and we gave little for the chances of the pilot getting out in time. Al Pinner, one of my pilots, started to run towards the crash until I told him not to – he would only have got in the way of the fire crews who were trained to deal with such an eventuality. Most fortunately, the aircraft had landed the right way up and the pilot had managed to get out before the fire started, suffering only a broken ankle, similar to Stan Witchall in the desert, and minor burns.

By the end of the year, the squadron was back up to a workable number of GR5s and, in January 1992, we deployed to RAF Valley for a missile practice camp firing AIM-9G Sidewinder air-to-air missiles in Cardigan Bay. Given the rocket motors on these missiles and the range they can travel, the 'safety trace' for such firings is huge and the exercise is very tightly controlled; besides any other consideration, each Sidewinder cost in the region of £50K. Blessed with good weather, we started our firings on Monday lunchtime and had fired our complete allocation of missiles by midday Tuesday.

During my first winter we deployed the whole squadron to Scotland for an operational training phase. My flight commanders told me that the squadron had done this for the last couple of years and persuaded me that it would provide excellent training, despite my concerns over the risk of limited flying in Scotland early in the year due to the weather. Having got my agreement, Gerry Humphreys, the flight commander with responsibility for organising the detachment, went off to book one of the Scottish bases for our two-week detachment. Somewhat concerned, he reappeared a few days later to say that none of Leuchars, Lossiemouth or Kinloss were able to accept us and the only RAF base north of the border that could accept us was Machrihanish on the very remote Kintyre Peninsular (well away from the social delights of St Andrews or even Elgin in the Moray Firth). "OK," I said, "since you told me the training value is that good, we'll go to Macrihanish." Being relatively new in command of the squadron, this decision had the potential to backfire on me with all squadron personnel if the detachment was not a success, especially on the social side.

We arrived on the afternoon of a cold, wet, gloomy February day and, with the next day's flying organised, I went down to the officers' mess with my squadron leaders to put our bags in our rooms and have a beer. Having been given room numbers we were directed to go outside, into a pitch-black night and half a gale, round the back of the main mess building to some old World War II single-storey accommodation blocks. As the squadron commander, having been allocated a block with my squadron leaders, I thought we would probably have some reasonable rooms and facilities. How wrong could I have been! My door only opened halfway then came up against some furniture; I pushed my bags in and squeezed through after them, arriving in a tiny room that had a very basic metal-framed single bed, a wardrobe and a chest of drawers which left virtually no floor space. Going down the corridor to find the washing facilities (there were certainly none in the room), my squadron leaders had already got there and discovered that there was no hot water for a wash or a shower. By this time the whole situation was so farcical that we saw the funny side of it and went off for a much-needed beer. The next day we had hot water and we quickly got used to sleeping in our tiny cells.

The aims of the detachment were to qualify all the pilots down to 100 ft low flying, to carry out weapons delivery on the Scottish air-to-ground ranges (including 1,000-lb high-explosive bombs on Garvie Island) and to train with the Harrier's Zeus electronic warfare system against the Tornado F3 wing at RAF Leeming. This was all well achieved although not without one particular bit of excitement which I could well have done without.

First thing one morning I was programmed to lead a three-ship, each of us with 1,000-lb high-explosive bombs to drop on Garvie Island off Cape Wrath, on a specified TOT. It was unpleasantly cold and, being early morning in February, the light levels were low as we taxied out for take-off. Approaching the runway, we were given clearance to take off but, by this time, I had discovered a problem with my aircraft and decided it would not be wise to get airborne. As the only way for me to get back to the dispersal was to taxi down the runway, air traffic cleared me down the runway but failed to cancel our take-off clearance. In accordance with the standing operating procedures, I set off taxiing down the right-hand side of the runway (the clearing side). It was some way to the turn off to exit the runway but before I got there, and without any warning, a Harrier suddenly flew right over the top of my aircraft missing me by a matter of feet. As if this was not sufficiently alarming, another Harrier then came down my left-hand side very fast; once past me, it then reapplied power and took off ahead of me. The aircraft were, of course, my Nos. 2 and 3 who, unable to see me due to undulations in the runway, and not helped in seeing where I was by the low light levels, had assumed that I had cleared the runway. Having not had their take-off clearance cancelled, and now running short of time to make the TOT, they elected to take off. By the time my No. 2 saw me, he had no option but to firewall the throttle and take off over the top of me; fortunately he had just enough room to do this. (He could not go around me as he would have collided with No. 3 who was taking off on his wing.) As No. 2 applied full power and left him, No. 3 quickly realised what had happened, slammed his throttle closed, kept to the left-hand side of the runway avoiding me and, once past me, seeing that he had enough runway remaining, reapplied power and took off. Needless to say, there was a serious debrief of the whole incident with the three of us and with air traffic; a hugely embarrassing incident, and lessons learnt, which we shared with the whole squadron that evening. It could have ended so badly – and we had all had 1,000-lb high-explosive bombs on board.

Socially the only town anywhere near Machrihanish is Campbeltown, a very pleasant seaside town, but not a place that had much to offer an off-duty RAF squadron on detachment in the middle of winter. As a result, the onus was on the squadron to provide its own entertainment and I remember a very successful all-ranks quiz night and an officers vs senior NCOs games night which both went down well. I played golf with some of the other officers at the weekend and ultimately, much to my relief, the detachment was deemed a success by everyone, both socially and professionally.

We even had a visit from Syd Morris, the station commander at Wittering, who told me he would come up to Machrihanish to see how we were getting on. On the day of his visit, I planned to fly first thing so I could host Syd late morning and over lunch before I went flying again later in the day. However, by the time I got back on the ground from my first trip, Syd had arrived at Machrihanish in a T4 with OC Admin Wing in the back, he had stolen my staff car and gone off to play golf. I went flying as planned later in the day, during which time Syd had returned from golf, said hello to a few people, got back in the T4 and disappeared back to Wittering, so I never did see him at Machrihanish.

At this time in the early 1990s, following the fall of the Berlin Wall and the breaking-up of the USSR which signalled the end of the Cold War, it was not clear what threat we were now training for. Whilst the squadron remained assigned to the flanks of NATO, we could continue to practise and train for this now unlikely threat (as we did over the coming years), but I felt that the Harrier Force should maintain a capability for off-base operations which had over many years been refined to a fine art in Germany. There were two reasons for maintaining this capability: firstly, if we could operate off-base, anything else we could be asked to do such as a 'bare base' deployment would be relatively easy by comparison; and secondly, maintaining an off-base capability relies on experienced personnel available to lead and train the newer personnel and once those experienced people have gone, it would be very difficult to regenerate the capability. I therefore felt we should try to maintain the capability whilst we still had people experienced in off-base operations in the Harrier Force.

Consequently, given that we had an excellent Harrier field training site at Wittering off the far end of the airfield (Vigo Wood), it was agreed that we would carry out a training exercise there in April. Once this was planned, HQ Strike Command declared that they wished to carry out an OPEVAL (a national operational evaluation of the squadron, as opposed to a NATO TACEVAL) in May. The one-week training exercise went ahead in April and provided a very steep learning curve for many on the squadron, both pilots and ground crew, who had never been on a fieldex before. The weather was not kind, which added to the difficulties, including a number of aircraft bog-ins as some pilots failed to stay on the metal planking laid on the grass. Nevertheless, after a week in the field, much had been learnt or re-learnt about Harrier field operations and, as in Germany, such an undertaking (often in adverse conditions) helped to mould the squadron together as a team.

Without a Harrier site office, as existed in Germany, it was left to the squadron to decide on an off-base location for our OPEVAL and for us to do much of the planning of the site. Having reviewed a number of possible options, we decided that RAF Kemble, operating from a large wood on the north side of the airfield, would provide us with a suitable venue for the deployment (Exercise Mayfly). My squadron engineers and the logisticians at Wittering did a great job of getting all the equipment and convoys organised and in early May the whole squadron deployed to Kemble ably supported by RAF Wittering station personnel (some more willing than others), the Royal Engineers, the Tactical Supply Wing (aircraft fuel), the Mobile Catering Support Unit and the Mobile Meteorological Unit. Due to personal circumstances, I was unable to deploy on the day of the move and it was left to my executive officer,

Ashley Stevenson, and senior engineering officer, Peter Coyle, to set up the site which they did excellently.

The deployment was run in the same way as a Germany field exercise: week one was for training and week two was for the evaluation. We flew combat training sorties on the Saturday morning, enjoyed a bit of R&R on Saturday afternoon and Sunday and then went into the assessed exercise refreshed and ready on Monday morning of week two. On the Sunday afternoon I briefed the OPEVAL team on the site layout, safety issues and limitations on our operations; I also advised them that we did have a bare step-up site available and prepared (on the other side of the airfield) if our site should become compromised by the enemy during the exercise. I was very proud of what the squadron and supporting units had achieved in establishing and operating what was, to the best of my knowledge, the largest single Harrier site ever put into the field. I told all site personnel this at the final briefing on the Sunday afternoon before the exercise started. We had some 320 personnel fully equipped and ready to operate ten Harrier GR5s off-base and with the ability to continue to operate following nuclear, chemical and biological (NBC) attacks. I was confident that we would give a good account of ourselves.

The weather was kind for the evaluation phase of the exercise (if anything too kind temperature-wise) and from my perspective all appeared to go well. The tasks were fed through to the pilots who flew the missions as required; the engineers prepared and despatched the aircraft without any major issues; and ground incidents were handled competently. And all of this was done during escalating exercise tension culminating, inevitably, in the donning of, and operating in, NBC clothing and masks. On day three of the evaluation, the site was 'compromised' and, having despatched the aircraft, the rest of us made our way, commendably quickly, to the step-up site on the far side of the Kemble airfield to continue operations. Shortly after this, the exercise was terminated. The combination of high temperatures and NBC clothing did lead to one of my stalwart armament SNCOs, who had been giving it 100 per cent and leading his team from the front, having to be treated for dehydration. I was also aware that there had been something of an altercation between one of the OPEVAL evaluators and some of my engineers. The evaluator, who was an engineer, had taken it upon himself to start 'nit picking' over the state of one of the aircraft's F700 (technical log). My engineers, quite rightly, were of the view that this had nothing to do with an *operational* evaluation of the squadron and told the evaluator as much. These two issues aside and, with the benefit of a number of Harrier MAXEVALs and TACEVALs behind me on 3(F) Squadron in Germany as a site commander, by the end of the exercise I was fully expecting good assessments for the squadron's performance. That evening, I went off to a local Cotswold pub with my flight commanders and the senior engineering officer (SEngO) to relax and to congratulate ourselves on a difficult job done well – or so we thought.

A few days later, back at Wittering, we met with the OPEVAL team chief to receive the evaluation debrief. Whilst the marks were generally satisfactory with some marginal areas, they were a huge disappointment. I did not feel that they reflected the magnitude of the task the squadron and supporting units had undertaken, nor did they give a true account of the performance that we produced on the day. There was a general feeling that the assessment had been done as a comparison against on-base operations, which was all the evaluators would have known. I knew from my time in Germany that we had been receiving 1s on TACEVAL (the highest rating) for a performance that was certainly no better than what we had done at Kemble.

An amusing sequel to this ultimately rather disappointing outcome was that a few weeks later we had a visit to the squadron by the Air Officer Commanding No. 1 Group, AVM Dick Johns, who had been the station commander and Harrier Force commander whilst I had been at Gütersloh. The AOC had not said anything to me about the OPEVAL but when we walked into the hangar to meet some of the ground crew, he saw one of his old engineers from when he had been OC 3(F) Squadron a number of years back. Once they had exchanged pleasantries, without any prompting from me or anyone else, the chief technician laid into the AOC telling him that the OPEVAL grades were a gross injustice to the squadron as what we had done at Kemble was as good, if not better, than anything that had been done during the whole of his time in Germany. Whatever the truth of the matter, I never received any adverse feedback from those above me in the command chain over the OPEVAL and it would appear not to have had any effect on my subsequent promotion.

Back at Wittering in June 1992, we began receiving the first Harrier GR7s, which we would use to assess the feasibility of introducing night operations to the Harrier Force front line; this was to be the squadron's main focus over the next two years. The GR7 was the night-capable version of the GR5, which incorporated forward-looking infrared (FLIR), cockpit lighting compatible with night-vision goggles (NVGs) and a digital moving map display. In due course we also had GPS retro-fitted which, linked to the inertial system, gave us excellent navigational accuracy.

THE INTRODUCTION OF NIGHT OPERATIONS TO THE HARRIER FRONT LINE

No. 1(Fighter) Squadron 1992–1994 – The Squadron Commander's Perspective

This account was originally published in *Harrier Boys Volume Two* by Bob Marston. It has been amended slightly.

With three tours on the Harrier behind me, I was only too well aware of the demands of daylight single-seat offensive-support operations – and now we were being asked to do this at night! Up until this time, Harriers and night flying had been very rare bed-fellows. I recalled a time in the Falkland Islands early in 1983 when 'normality' began to return to RAF Stanley and OC Supply Squadron put out a station order to the effect that stores exchanges and issues would only be available between the hours of 0900 and 1600. As Harrier Flight commander I promptly put out an order saying that with immediate effect the Harrier Detachment would only be available for the air defence of the Falkland Islands between 0900 and 1600 and that outside these times the Phantom Detachment should be contacted for assistance. This was not far from the truth, of course, since our role in that theatre was limited to daylight hours only. But now the Harrier Force was going operational in the dark.

Although by this stage in my RAF career I had about 130 hours' night flying, quite a lot for a Harrier pilot but much of it gained as a flying instructor, the rest of the squadron had much less night experience than me. In addition, from a personal perspective, I had now been away from flying for over five years on staff tours. So one concern was whether I would have the flying currency to lead the squadron into this task and another was whether we all had the experience and background to undertake the new role. But what did the task entail? And what experience and background did we need? My own first-hand experience was confined to one night trip in a Harrier T4 with Keith Grumbley when he was boss of the Strike Attack Operational Evaluation Unit (SAOEU); he had NVGs, I didn't, and it was frankly pretty scary. But there was no doubt that the ground-breaking work undertaken by the SAOEU, much of it on the very demanding Nightrider[19], had provided a sound experiential basis for us to introduce night operations to the front line and had established a firm belief that a Harrier day combat-ready pilot should be able to cope with the demands of this new role.

In the event, due to delays with the introduction of the GR7, the night programme was delayed for a year and it was not until September 1992 that the squadron undertook the first night sorties. Even then, although the NVG compatible cockpit lighting was ready, the goggle ejection system[20] was not, and we commenced night operations without the benefit of NVGs. By this time I had acquired an OC Night – Mike Harwood – from the SAOEU, who was experienced in NVG/FLIR[21] operations and who would be key to converting the squadron to night operations over the coming two night seasons. Since none of us apart from Mike had any experience of NVG/FLIR, it was incumbent on me as the squadron commander to be at the forefront of whatever we were doing. Whilst Mike had devised a night combat-ready work-up syllabus, I would have to ensure that I was happy with what we were asking the pilots to do and put a stop to anything I was uncertain about. I chose Rob Adlam, my squadron QWI, an experienced and very capable Harrier operator, as the man to join me in stepping out into dark and unknown territory and to help me decide what was safe.

Without the NVG clearance, Mike's view was that we would benefit from flying without the goggles (using the naked eye and FLIR only) as this would help build our confidence. The early sorties achieved this although, at times, they could be somewhat exciting. Rob and I did quite a lot of night close formation and although the exterior night formation panels[22] on the aircraft were excellent, without NVGs it was very difficult to assess how close you were to the other aircraft. I remember one night Rob calmly asking me to ease out as I had settled in too close to his aircraft without realising it – he could feel me interfering with the airflow over his aircraft. Trust was certainly required. Night-bombing at 150 feet at Tain, rejoins into close formation and night air-to-air refuelling all had their moments as well. We quickly realised that we were all learning things by experience and, at times, making mistakes that we might not want to own up to. So we introduced a 'lessons learnt'/ honesty book in which all pilots could put down anything that they thought might be of use to the rest of the team; entries were unattributable and hugely valuable.

It was a great source of comfort to me that I had one pilot who had a good amount of experience of electro-optical (EO) operations; I also knew that Mike had an excellent reputation from his time with the Strike Attack Operational Evaluation Unit. However, this did not assuage my concern that, as squadron commander, I was responsible for the two-year trial and that no-one above me in the command chain had any appreciation of what we would be doing. It was a most unusual but privileged position to be in and I was very grateful for the trust placed in me by the Air Officer Commanding 1 Group, his staff and both my station commanders over this period; they allowed us to get on with the job without any undue interference. I remember Group Captain Syd Morris saying that he always knew when we were getting towards the end of a night-training period because I would start getting tetchy. It wasn't a complaint just a statement of fact that I, and the rest of the squadron pilots, would get noticeably tired after two to three weeks of continuous night operations.

In November 1992 we got the NVG clearance and began the night work-up properly. By this time it had become apparent that the aircraft modification programme for the blow-off system was going to be protracted, which would limit the pilot conversion rate; this presented a new challenge since all the pilots were desperate to be at the cutting edge of bringing night operations into front-line service. Reluctantly, Mike and I agreed that we would have to split the squadron into three groups of pilots: the A team who would get priority for night assets this first winter; the B team who would convert at a slower rate; and the non-combat-ready pilots who would not commence night operations until they were day combat ready (they would then join the B team). The exec, Ashley Stevenson, took over leadership of the B team. Unfortunately, but quite understandably, the B team always felt like second-class citizens. I was acutely aware of this and had trouble addressing it to their satisfaction. However, it says much for the commitment and engagement of the pilots on the squadron that this was how they felt. With the squadron operating a day team

First night of electro-optical flying with Rob Adlam, RAF Wittering, 17 November 1992.

and a night team during our two or three-week night phases, just carrying out routine but important tasks like execs' meetings could prove difficult. At one point, it had got so problematic to get the flight commanders and SEngO together that I said we would have to meet after night flying one night; and this we did, starting a meeting at 0030 in the boss's office with a beer. In addition, with a pilot complement of just 16, all the execs (including the boss) took their turn as duty authorising officer, which doubled up as duty pilot in the tower once the last night wave had been despatched.

Late in 1992 we were fully into night operations, with Mike selecting the night-flying weeks depending on the phases of the moon, i.e. the light available, initially flying on well-lit nights but later on choosing nights with no moon at all so that we were using two millilux[23] from the stars as best to find our way around at 250 feet. During our night-flying weeks, the squadron and the station had to adjust to us starting flying in the early afternoon and continuing through until around 11pm when the night low-level flying system closed. Again, I was very fortunate to have two very understanding and supportive station commanders throughout this period. Besides reorganising the station's support to meet our requirements, they also agreed to keep the officers' mess bar open late for us so we could unwind after flying. It was not unusual for pilots to find it difficult to sleep after night operations because of the level of adrenalin in the body;

unsurprisingly perhaps, we found that a beer or two with the other night fliers was a good way to wind down.

We had some visitors to the squadron to see what we were doing. I took a film crew from ITV to the ATC tower to 'watch' a pair of aircraft departing on a night sortie. They were somewhat surprised as we drove to the tower when I explained that I had 12 £14 million Harrier GR7s on my squadron, 130 personnel and the task of introducing night operations to the Harrier Force, yet my squadron commander's car, which they were travelling in, was so unroadworthy that it was not allowed off-base! In the tower we could hear the engines start and listened to the leader call for taxi clearance. We then went out onto the balcony to look across a completely black airfield (no light from the squadron apron either as the ground crew had, as always, despatched the aircraft in blackout conditions) and saw just two red wing lights accelerate rapidly across the unlit airfield, lift off and depart at low level towards the west; and that was it. We also had a visit from a very senior RAF officer from the Ministry of Defence who came to find out what we were doing. Unfortunately, he impressed in completely the wrong way by interrupting the two ex-Tornado Gulf War pilots, who were debriefing their mission for his benefit, to complain that they were not wearing the aircrew watches which he had fought a major battle in the MoD to acquire. My pilots wrapped up the debrief very quickly after that somewhat misplaced interjection.

Whilst the focus of what was happening on the squadron was decidedly on air operations, it should also be remembered that the advent of night operations meant a huge change for the ground crew as well. In between night seasons, the squadron also deployed to Incirlik, Turkey, on Operation Warden, carrying out reconnaissance missions into northern Iraq in the aftermath of the First Gulf War. With Peter Coyle, followed by Peter Ewen as SEngO, and 'Charlie' Chaplin as squadron warrant officer, everything that we asked of the ground crew was done without quibble. Despatching the aircraft from a fully dark apron became routine and this continued whenever possible on deployed operations as well. It was probably just as well that the Health and Safety at Work rep tended not to come out at night.

Towards the end of the first night season, we deployed to RAF Leuchars in February 1993 to carry out night operations in the Highlands, including dropping 1,000-lb high-explosive bombs on Garvie Island off Cape Wrath. Late one afternoon, some Tornado F3 crews came into the planning room and asked if we were planning our low level for the next day, but thought we were joking when we said we were going to fly the route that night in the dark. I recall one particular mission on that detachment when I was flying as No. 2 to Chris 'Snorts' Norton, one of the most capable first tourists in the Harrier Force at the time. Encountering some decidedly poor weather, I eventually gave up the struggle to keep with him in fighting wing[24] as we went through rain and snow showers in some demanding terrain and I pulled up above cloud. A quick exchange on the radio and we had agreed a rendezvous en route/on-time with Snorts taking the north side of a valley whilst I let down through gaps in five octas[25] of cloud cover on the south side of the valley. The GPS/INS[26] navigational display, software and NVG/FLIR combination in the GR7 made such a task reasonably straightforward. At the debrief I asked Snorts how he had managed to remain at low level when I had pulled out? His answer was that as the NVG performance reduced, you could get more texture from (i.e. visibility of) the terrain by flying lower. I could appreciate that he was technically correct, but we all had to decide where our own personal levels of ability, comfort and safety lay.

A couple of months later the squadron deployed to the USMC base at Yuma, Arizona, to undertake night operations in the desert (very low or nil cultural lighting/poor NVG – dry air/excellent FLIR). A year later, at the end of the second night season, we carried out night operations out of Bardufoss in northern Norway in March (some good cultural lighting with the Northern Lights and white snow-covered ground – moist air/limited FLIR). It was during this exercise that Gerry Humphreys and I undertook some of the first night close air support profiles as a pair, inputting the FAC brief direct into the GR7 through the upfront controller[27] as writing the brief down in the cockpit at night was a non-starter. Towards the end of the Bardufoss detachment we were hosting a cocktail party for our Norwegian hosts, which I required all officers to attend. Mike Harwood insisted that that night was the last opportunity in the season for Ian MacDonald, the next OC Night, to get night-combat ready. I agreed that they could fly the mission provided they overflew the officers' mess at exactly 2000hrs at 250 feet as part of Ian's combat-ready check. I started my words of thanks to our hosts at 1957 and at 1959:30, after a timing nod from Gerry Humphreys, made our apologies for the two missing officers – "Oh, here they come now!" – right on cue. The head-up display video was impressive and Ian passed his combat-ready check. In Norway we flew our first night three-ship and on return to Wittering, the last night trip of the season, and my last-ever Harrier flight, was the first night four-ship. The next day I handed over command of the squadron to David Walker who would shortly lead the squadron into night operations for real in Bosnia.

Looking back, it was an amazing period – an exciting, challenging yet hugely rewarding 18 months. We had progressed from a bunch of novices who knew nothing about offensive night operations, to a team that could fly tactically at night with up to four aircraft, in different EO conditions, evade an aggressive air threat, locate a target and deliver weapons with a high level of precision. But what of the risks? I believe that by acknowledging from the outset that what we were about to undertake was inherently risky, we had already gone a long way to reducing the risk. In addition, by being open and honest when we could, using the honesty book when we couldn't, and carrying out extended debriefs in the bar into the small hours of the morning when necessary, we shared our experiences for the benefit of all. We had a huge respect for the role we were trying to get to grips with and this, as much as anything else, helped keep us safe. Tragically, we did lose a squadron pilot over this period, but not at night. We had asked for the US Marine Corps exchange post to be transferred from IV(AC) Squadron for a tour so that we could have an experienced AV-8B night operator and benefit from the USMC's night experience. Very sadly, Captain Brenden Hearney was killed when he flew into the ground on a low-level training exercise whilst he was attached to No. 233 Operational Conversion Unit in January 1994.

There will not be a pilot who flew with the squadron over the two night seasons that did not have 'moments' and, in truth, we were fortunate that none of them became more than that. I will end by sharing two of mine, which typify the pitfalls and risks:

Rejoining the circuit at Yuma, Arizona as leader with my No. 2 in fighting wing. The airfield is fully lit so we have removed our NVGs[28] and my No. 2 is now following my tail light and using air-to-air TACAN to maintain separation from me. I decide that I am too close to the airfield so I turn left by about 40 degrees to give us some more separation from the runway before starting a turn back towards the airfield. Shortly after rolling out of the turn there is an incredibly loud roar as my No. 2 flies right over the top of my aircraft, missing me by a matter of feet. He had not detected my turn and we were only saved from a mid-air collision by his religiously sticking to always staying high on the leader in fighting wing when close to the ground.

A singleton low-level night-navigation exercise. I coast out from the Lake District heading west over a very dark sea climbing to 1,500 feet. There is virtually no goggle performance and little on the FLIR. I start my turn back to coast in again and begin a gentle descent on instruments so that I will be around 500 feet crossing the coast. The trip has gone well and I am feeling on top of things, so when I realise I have to do something in the cockpit I start to do it. A small voice in the back of my mind is telling me that this is not a good time but I assure myself it will only take a moment. Before I know it, the radio altimeter low-altitude warner is beeping (225 feet) and I am rolling wings level, pulling hard back on the stick and seeing 180 feet on the altimeter. A salutary reminder – and I sadly needed one – that there is no room for complacency or over-confidence in single-seat, night, low-level operations.

We shared an amazing experience on No. 1(Fighter) Squadron over that 18-month period. The most gratifying part was that the squadron's two-season night trial, built upon the excellent work of the SAOEU, provided a sound basis for night operations to be adopted by the other squadrons. This enabled night EO operations to be used in anger in a number of operational theatres by the Harrier Force, from 1994 until its demise in 2010.

Another event that I remember well from night flying was joining a tanker for my first night air-to-air refuelling sortie. We would use our NVGs but, because of the floodlighting on the underside of the tanker, we would discard our goggles whilst engaged in the refuelling. I was being vectored onto the tanker by a ground-fighter controller and, as usual, was being held at 2,000 feet vertical separation below to preclude any risk of collision. As I commenced the turn to roll out behind the VC10 tanker I was in cloud, relying totally on my flight instruments, with an occasional look out of the cockpit for any sign of the tanker. Halfway round the turn, the fighter controller called the VC10's range from me as five miles. At this point, I came out of cloud and saw the tanker right on my nose and, apparently, very very close. My instinct was to slam the throttle closed and put the airbrakes out. Overcoming this instinct, I quickly rationalised that I was 2,000 feet separated, I was five miles from the VC10 (my air-to-air TACAN confirmed this) and the tanker only appeared very close because the NVGs were reacting so powerfully to the refuelling lighting underneath the tanker. The goggles were truly amazing bits of kit.

Returning to day operations, one flight I particularly recall was on the ACMI range in Sardinia. I was scheduled to fly as No. 2 to one of my junior pilots against four RAF Phantoms. The brief was that we would approach the F-4s head-on at 35,000 feet in very wide, four-mile spread, formation and as we approached the F-4's missile launch range (20 miles) we would roll to the inverted, pull down into a very steep dive and pull out at 10,000 feet; hopefully this would present the fighters with a serious tracking problem as they came up to missile launch. As we climbed up to 35,000 feet, I could see that the game plan was going to be difficult to deliver. Typically for the ACMI range, which was entirely over the very deep blue Mediterranean Sea, there was a poor horizon due to haze and the sky above was very blue as well – typical 'goldfish bowl' conditions in which it was quite possible to get disorientated despite it being broad daylight. I was having trouble keeping my leader sighted as we widened to four miles on the run-in towards the fighters and when we came to execute the plan, I rolled to the inverted, started the pull down, established visual with my leader, took my eyes off him for a moment to check the flight instruments to stay orientated, looked back for my leader and couldn't see him at all. I then spent too long trying to find him visually and then realised I was, in fact, totally disorientated. By the time I had sorted myself out, I was below 10,000 feet with no idea where my leader was and, anyway, by this time the F-4s had claimed a shot on me. I have no idea how fast I went during the recovery from my disorientation but it was almost certainly the fastest I ever went in the Harrier (and probably the only time I went supersonic in one).

Throughout my tour we retained a T4 as a squadron hack, which was very useful and we often used this aircraft as an offensive air asset (bounce) against our GR5/7s. Given the obvious differences between the T4 and the newer marks, and the safety issues as previously discussed, we restricted flying this aircraft to those of us with a lot of experience on the T4/GR3. On occasions, I sometimes found myself flying the aircraft at the limits of the required 30-day currency. Whilst I had no issues with the technicalities of climbing back into the T4, after some weeks away from it the cramped cockpit, poor instrumentation and ergonomics, and abysmal visibility compared with the GR5/7 always came as a bit of a surprise for the first 20 minutes or so; thereafter I always found it a delight to be back in the old jet. Given the vagaries of the inertial navigation system, acting as bounce was always testing once you had got used to relying on the excellent navigational accuracy of the GR7.

There are a number of events that stand out well in my memory from the middle of my tour in command of the squadron. In June 1992 we celebrated the squadron's 80th birthday over a weekend with a great turnout of past members, including Ned Crowley (Hurricanes in 1941), who performed his party trick of drinking half a pint of beer whilst standing on his head; he was then aged 73. The squadron flew a 'Flying One' of Harriers, the Red Arrows displayed, we had a sunset ceremony, and a church service on Sunday morning followed by a jazz lunch in the garden of the officers' mess. The whole weekend was organised with a great deal of style by Mike Howes. That summer, Flight Lieutenant Chris Huckstep had to eject on take-off following an engine failure, fortunately he was unscathed.

I also recall standing in front of air traffic at Wittering the following year with Pete Day, Syd Morris's replacement as station commander, who had been the executive officer on the squadron in July 1977, to watch the arrival of three Harrier GR3s from Belize as the commitment ended. The formation leader was John Finlayson, the last flight commander in Belize, and he had been a flight commander on the squadron during the reinforcement in 1977. It was the end of a small chapter in the history of the RAF and the Harrier Force that brought three of us original participants back together at Wittering, the 'Home of the Harrier', where it all started.

My wife Jill had been diagnosed with cancer during our last year in Gloucester, which had necessitated surgery and chemotherapy. After some time in remission, tragically the cancer spread further during the first half of my tour as OC 1(F) Squadron and she very sadly passed away in March 1993. The RAF offered me the option of moving to the OCU in command, in many ways a less-demanding job which would not involve any detachments. Whilst I was grateful for their consideration, I was very reluctant to walk away from the squadron at this point, not least as it would have been very difficult for the new incumbent to take on the night trial which was now halfway through. I therefore turned the offer down and stayed with the squadron, deploying to Yuma in Arizona for the culmination of the first night season a few weeks after Jill's death.

I had an unusual and somewhat concerning experience on this detachment. One evening, having already flown one night mission, I was not expecting to fly again when Mike Harwood came up to me and said there was a spare jet and I could get a singleton trip in if I had a route ready to go, which I did. I quickly got myself sorted, got out to the jet in good time and entered the route data into the aircraft manually (as we had to at this time). The aircraft inertial system now knew my whole route and provided the inertial remained accurate, I could follow the aircraft navigation system with confidence. Since our night-flying jets had the GPS linked up to the inertial, provided a green light stayed lit in the cockpit, we knew that the inertial was being continually updated by the GPS to keep it very accurate. Having been operating this system all winter we knew it worked well and had lots of confidence in it. After an uneventful take-off, I entered the low-level route flying at 250 feet and 420 knots. There was little to navigate on crossing the desert in Arizona but the green light was on so all was well. As I approached the first IP (initial point), a small road junction prior to the first target, I was not too concerned not to see it and pressed on to the target now at 480 knots. As I approached the target area, the terrain did not look right and I could not see anything resembling the target (I don't remember what it was but I should have been able to see it). The green light was still there but I was now very concerned that I was not where I thought I was (or at least where the aircraft's navigation system thought I was). At this point I pulled up and climbed to 4,000 feet before getting a lock-on to the TACAN navigational beacon at Yuma. This confirmed that I was some way from where I should be. I aborted the mission and by the time I got back on the ground the aircraft had a navigational error in excess of 120 miles although the green light was still lit.

Besides the runway incident at Macrihanish, the near mid-air at Yuma and the close shave with the sea off the Lake District, there was one other incident I had whilst in command of the squadron which could also have had a bad outcome. One afternoon, I was leading one of my junior pilots on a low-level training mission into Wales. Flying across the Cotswold Edge near Broadway Tower, I saw a bird pass very close down the side of the aircraft and thought it might have gone into the engine. Although everything appeared to be OK with the aircraft, I elected to divert into RAF Brize Norton to check the aircraft over. I told my No. 2 to continue the mission on his own. Having dumped some fuel to get down to landing weight, I made an uneventful landing at Brize. A careful check of the engine and airframe revealed no damage so I decided to take on some fuel and continue with the mission as planned on my own. It was late afternoon as I headed north through mid Wales at 250 feet. Whilst flying a single aircraft on a training flight was quite normal, being part of a formation of two or more aircraft obviously improved the lookout for other aircraft. This was especially useful at low level where most other military low-level aircraft were flying at around the same height, training to use the terrain to remain shielded from ground-based and airborne radars. Always mindful of the risk of collision, I was keeping a continuous visual scan around me for other aircraft when I was suddenly aware of a movement close to my right. Instinctively I pulled hard on the control column to pitch the aircraft rapidly up then banked to the right to see a Tornado GR1/4 fly straight underneath me. There was no reaction from the Tornado – they hadn't acknowledged my presence with a wing waggle – so it was apparent that they had failed to see me. There is little doubt in my mind that if I had not acted instinctively to take avoidance action on the movement I perceived, at best we would have passed very, very close to each other.

In July 1993 I was promoted to group captain and was due to leave the squadron at the end of the summer on completion of my two-and-a-half-year tour. However, with the night trial still having its second winter ahead, it was not a good time for a new squadron commander, who would not have any night experience, to take over. Consequently, I asked to remain with the squadron to see the night trial through until spring 1994. This also meant that I flew missions into northern Iraq under Operation Warden in the late summer of 1993.

I deployed to Incirlik in Turkey via Brize Norton. Positioning the night before the flight out on a VC10, I met up with OC 101 Squadron, Wing Commander Pete Scoffham, in the bar at Gateway House. He was deploying to Incirlik to take over the VC10 refuelling detachment and I had last seen him at Cranwell when he graduated two years ahead of me. We were both in civvies and had a couple of beers and a chat before going off to bed. In the morning I went into breakfast in uniform and saw Pete who did a double take and said, "You're a bloody group captain!" He couldn't get over the fact that I had left Cranwell two years after him but was now promoted to group captain ahead of him. He kept going up to his crew members and saying, "How did OC 1 Squadron get to group captain ahead of me?!" It was all in good humour and he said I would have to go flying with him whilst we were in Incirlik, which I did. On the day, his crew told me the boss himself had been down to the US Base Exchange to do the shopping for our 'special lunch'. As soon as we were airborne, Pete disappeared down the back to the galley leaving me in the captain's seat whilst he prepared lunch. I chatted with the rest of the crew (including being 'chatted up' by Pete's very attractive lady navigator, with the rest of the crew listening in on intercom – a put-up job for sure) and watched my Harriers refuelling from the VC10 before they set off on their recce task. The starter arrived as we transited out to Iraq, we enjoyed an excellent steak with all the trimmings whilst we were on task in northern Iraqi airspace and finished with dessert and coffee en route back to Incirlik. The 'chef' made it back into the captain's seat shortly before landing. It was a great day out and I think Pete got his longed-for promotion shortly after.

AIR OPERATIONS – IRAQ 1993

It is very hot. It is summer in Turkey. The First Gulf War took place three years ago. Little appreciated by the general public in the United Kingdom, the airspace over the north and south of Iraq remain operational theatres for the Royal Air Force. Under UN mandates, allied air operations are still being launched to prevent the Iraqis flying into these areas; this will help protect the Marsh Arabs in the south and the Kurds in the north. We are flying our Harriers out of Incirlik in Turkey into the northern no-fly zone in a coalition with the United States Air Force and the French air force.

The aircraft I am now flying is the Harrier GR7. This is a completely different beast from the old GR3. It has a bigger wing which gives three major benefits: much more fuel (so a greater range), a greater weapon load, and a hugely improved turning performance for air combat. In addition, the aircraft has an excellent navigational and weapon-aiming system and it is equipped for night operations, employing night vision goggles and forward-looking infrared systems. Our planned operations into northern Iraq are day only, so we will not require the night capability, but the additional range, weapon load and navigational accuracy of our new aircraft will be of great benefit. Our main task will be to carry out reconnaissance of Saddam's troop dispositions and missile systems using our photographic reconnaissance pod. But the aircraft will always carry two Sidewinder AIM-9L air-to-air missiles for self-defence and, on occasions, we will also carry a full live bomb load. Although I am now the squadron commander, and have overall responsibility for everything the squadron does in this operational theatre, on the detachment to Turkey I will fly as regularly as any other squadron pilot. I am paired-up for operations with our American exchange officer, Joe Andrews. We will fly all the missions in the coming weeks together in two of our Harrier GR7s. Flying together we can give each other mutual support in a physical sense against any hostile threat. We will also be able to provide some measure of psychological support as we fly our single-seat, single-engine aircraft over hostile areas for one to two hours on each mission.

This morning, coalition aircraft over northern Iraq reported the launch of an Iraqi surface-to-air missile (SAM) at allied aircraft. USAF F-16 fighter aircraft were called in to drop live munitions on the launch site immediately and it is now uncertain what the further response will be. Our Harriers are due to take part in this afternoon's operation: it may be cancelled; it may go as planned, as a tactical reconnaissance; or it may be launched as an offensive (bombing) operation. Earlier I went down to the Combined Operations Centre (COC) with the RAF Detachment commander, a fellow group captain, where the response to the morning's events was bullish. I left there to plan the mission as tasked and shortly before we are due to walk out to the aircraft, it is confirmed that the mission will go as planned to carry out reconnaissance, not as a retaliatory strike.

Planning and briefing complete between us, Joe and I go to the intelligence section for an update. This confirms what we have been told earlier. We then go through the process of ensuring that all our personal items have been left behind. In the event of our ejecting from the aircraft and falling into the hands of the Iraqis, they could use any personal information to undermine our resistance to questioning following capture. We are then issued with our 'combat vest'. This contains our personal weapon (with live rounds and a spare clip of ammunition), gold florins (to buy our way out of trouble), a handheld GPS system and radio (for rescue purposes) and additional items of personal survival gear. We are now ready to make our way to the aircraft – and shoot or buy our way out of trouble in northern Iraq.

When I carried out my training in the RAF, we were sent on Exercise King Rock in Germany, which tested our ability to survive following a parachute descent into enemy territory. Our experiences over the four-day 'escape and evasion' phase provided a certain amount of hilarity, but there was also a reality to that training 23 years ago which helped prepare me for the possibilities of the situation I now face. If I eject over the border, I am confident that the coalition will endeavour to rescue me without delay. If they cannot get me out, I will have to try and avoid capture for as long as possible.

We go to sign for our aircraft. This involves completing the Form 700, the technical log for the aircraft. In signing this document, I am accepting responsibility for the aircraft until we return from our mission and I hand the aircraft back to the ground crew. As I finish signing the Form 700, I realise that, unusually, all the squadron ground crew are there. The events of the morning (the SAM launch and the subsequent bombing by the F-16s) have been widely reported, but the ground crew have no idea what is happening now. I have been so involved in visiting the COC and preparing for the mission that I have not thought to get word to them. Impromptu, I give them a brief rundown of what has happened and what our mission will be this afternoon. The bottom line is that we have no idea what the Iraqi response will be to the coalition's retaliation this morning.

I walk off to my aircraft alone, weighed down by the combat vest and life jacket, carrying flying helmet, maps and water for the mission. The sky is clear, the sun is high and the temperature is +36°C.

Cold War warriors would mostly agree that Her Majesty's Government were very reticent about handing out medals and it was notable that on Operation Warden our USAF colleagues received two medals for their participation whereas the RAF received none (not that I would suggest that we should go as far as following the American example). However, it was some five years later in about 1998 that, purely by chance, I happened to see a Defence Council instruction stating that it had been decided that those who had taken part in Operation Warden could apply for the award of a general service medal (GSM) – and this was for an operation that had been running since 1991.

In the autumn of 1993, I received a phone call from an army major in charge of a company of Territorial Army (TA) Royal Engineers whose role included Harrier support. He said that they would be building a Harrier field site on the old RAF airfield at Waterbeach near Cambridge and asked if there was any chance I could send a Harrier in to land on their pad; this would be an excellent reward for his men to see their hard work actually put to use. Being TA, the work would be done over a weekend, but they said they would be able to take an aircraft at lunchtime on the Saturday. Pete Day, now the station commander at Wittering, agreed that the airfield would open for this so I phoned the major back and said we would come, not with one but with four aircraft which he was delighted with. I did add, however, that it would be dependent on us getting a good lunch when we flew in.

On 27 November, I led three volunteer pilots (bribed into giving up their Saturdays on the promise of a good lunch) into Waterbeach where we all landed uneventfully on the newly laid pad and taxied into the hides. Once we had said hello to the army and sorted out the aircraft for the return flight to Wittering, the major told us that lunch was ready, started handing out mess tins and said we should head off to the field kitchen. Seeing the consternation on our faces, his men then drew back the flaps of a large mess tent to reveal tables set up for lunch with table cloths and all the trimmings. After a splendid lunch it was back into the cockpit for the flight home. Ashley Stevenson was site commander on the ground and had surveyed the take-off strip which we had visited before getting back into the jets. It looked pretty short to me, with small trees at the end, although we did not have much fuel on board and Ashley had calculated there was sufficient margin. I was first off and was still thinking it looked short as I hit full power and accelerated very rapidly towards the trees. Lifting off very close to the end, I rotated rapidly and brushed through the tops of the trees as I climbed away. Seeing my departure, the other three pilots wisely gave themselves a bit more room for their departures.

During my time on 1(F) Squadron I was blessed with some very good flight commanders. Chris 'Wedgie' Benn, who had been on 101 Entry at Cranwell a year behind me, was the exec when I arrived (he had also been my deputy flight commander in Germany). With great organisational skills, Wedgie was a huge assistance to me when I first arrived on the squadron. Les Garside-Beattie, Ashley Stevenson, Gerry Humphreys, Mike Harwood and Mark Green were the other flight commanders at various times for the main part of my tour. Due to posting dates, Ashley had the dubious honour of being my exec for an unusually long time, some 21 months. He was well respected across the

From left to right: Author, Squadron Leader Jim Mardon and Ashley Stevenson at Al Pinner's wedding. Ashley survived two ejections from Harriers, went on to command 3(F) Squadron and retired from the RAF as an air commodore.

squadron and did an excellent job of supporting me over this period even telling me quite rightly, on one occasion, to get out of the office and go home. Les, Gerry, Mike and Mark were all very capable flight commanders and each of them played a very full role in the performance and development of the squadron. As previously mentioned, I also had two excellent SEngOs, who ensured that the engineering side always supported the flying effort to the full. There is no doubt that the success of a tour as a squadron commander can hinge on the quality of the squadron leaders (and warrant officer and SNCOs) in support and, in this respect, I was very well served.

My 'swan song' before leaving the squadron was a four-ship southern ranger to Madrid for a weekend in January 1994. One of the potential problems with such flights was that we had no engineering support if anything went wrong with one of the aircraft whilst we were away. All went well on the route down (a single leg in the GR7 whereas we would have had to land and refuel in the GR3) and we had a very enjoyable weekend in Madrid. However, when we came to return to Wittering on the Monday morning, Mike Harwood found he had left the refuelling panel open on his jet. Unfortunately, doing this meant that the aircraft's refuelling circuits had been left powered up and the aircraft's battery was totally dead. It now transpired that Mike had to get back to the UK as he had a retired senior officer visiting him that evening and Gerry Humphreys had to get back as his wife was about to give birth, which left the boss and the junior pilot to sort the issue out. Mike and Gerry jumped into two of the aircraft and departed for the UK.

The language barrier with the Spanish air force engineers was the first hurdle but, even once that had been overcome, they couldn't suggest any solution – so much for NATO interoperability! We spoke with our own engineers back at base and they agreed our plan to start the auxiliary power unit (APU) with the good battery and then swap batteries to get the other aircraft started. Now the battery in the Harrier is in the rear bay and not easy to remove, especially when you haven't done it before. With the APU running, there is a huge amount of noise in the rear bay. Having started the APU, I then removed the good battery and replaced it with the discharged battery so it could charge up whilst we were still on the ground. I then took the good battery across to the other aircraft, put it into the rear compartment, connected it up then started the engine and we eventually got away back to Wittering.

Following detachments to Leuchars in February for operational low-flying (100 feet) and weapons training, and the following month to Bardufoss for night-flying training, my tour was nearing its end. On the evening of 28 April, I flew my last trip on the squadron, leading the Harrier Force's first night EO four-ship. The following day I handed over command to Wing Commander David Walker. I left the squadron one month shy of three years in command.

CHAPTER 10
STATION COMMANDER RAF SCAMPTON 1994–95

To be given command of any RAF station is a privilege and it is even more so when that station is one of the truly iconic stations of the RAF. Even today, more than 75 years after the event, many people are still aware of 617 Squadron, The Dam Busters, and the dams raid in May 1943 which was mounted from RAF Scampton. More recently the station had been a main operating base for the Vulcan but it was now the home of the Central Flying School, the Red Arrows, the Joint Arms Control Inspection Group (JACIG – a tri-service unit), and the Trade Management Training School (TMTS). I was scheduled to take over command of the station in September 1994 but, prior to that, I was required to undertake a number of courses to prepare me for this command.

On the flying side, this included undertaking instructor qualifications on the Tucano and on the Bulldog, two types I had never flown before, and a refresher course on the Hawk. Consequently, I had a very enjoyable few months learning to fly and instruct on the Tucano and the Bulldog at Scampton, which also gave me the welcome opportunity to get to know many of the people there before taking command. In addition, I detached to RAF Valley for two weeks to carry out a Hawk refresher course. Since I was to have the Red Arrows at Scampton, this course included some of the training the team members carried out prior to starting their work-up training. These flights were low-level aerobatics over the airfield at Valley and practice engine failures from various positions at low level which were carried out at the airfield at Mona. The latter flight was to enable team members to explore the capabilities of the Hawk in getting back onto the runway should they lose the engine during a display. Already a keen fan of the Hawk from my brief acquaintance with it at Brawdy a few years earlier, this flight confirmed the outstanding gliding and manoeuvring capabilities of this excellent aircraft.

I remember very well one particular flight on the Tucano with Dave Turner whilst I was doing my training. The Tucano was the only aircraft I have flown that was cleared for inverted spinning, consequently it was necessary to be competent at this to be signed off as a QFI on the aircraft. Dave gave me a very full and clear brief. Getting the aircraft into an inverted spin was an art in itself: at 140 knots 20–25 per cent torque raise the nose to 25–30 degrees above the horizon, at 110 knots set the throttle to idle and apply full aileron to roll to wings level inverted, at 95 knots check ailerons neutral and apply full rudder in the direction of spin required and push the control column fully forward. In the Pilots Notes there is then a clear warning about the risk of inadvertently applying full nose-down trim whilst holding the control column fully forward during the spin. If this happens, the aircraft will tuck under on recovery with very high stick loads and the pilot may think the aircraft is not responding. Indeed this had happened in the earlier days of the Tucano and this had resulted in the crew ejecting. Dave covered all this and we were good to go. Once airborne, I got us into the inverted spin OK first time. The entry was quite disorientating but once settled in the spin everything was quite steady although you could not see the horizon due to the low nose attitude. The recovery drill for the inverted spin was the same as the erect spin (apart from moving the control column aft rather than forwards) and the spin stopped straightaway. At this point I could feel the aircraft starting to bunt (i.e. tuck in). I immediately hit the trim button and started pulling back hard on the control column and we recovered without any further excitement. Analysing what had happened afterwards, although I was fully aware of the risk of applying full nose-down trim in the spin and had (I thought) put my hands away from the trim button, I had still managed to make contact with the unguarded trim button. Fortunately, having read the Pilots Notes and had a good brief from Dave, I recognised the problem as soon as it happened. In fact, Dave in the back didn't even realise it had occurred and it was the last time I made that mistake.

In September I took over command of the station from my predecessor Group Captain Ted Edwards. I had been told by Syd Morris at Wittering that running a station was, in many respects,

Taking over command of RAF Scampton from Group Captain Ted Edwards, September 1994.

a more interesting and challenging job than running a Harrier squadron; he was correct. Scampton was a delight of a station to command with its breadth of units but with a firm focus on flying. Central Flying School and the Red Arrows came under the commandant Central Flying School who had his HQ at Scampton. I knew Air

Pumping Red Arrows diesel (for smoke generation), RAF Lyneham in June 1995.

Commodore Simon Bostock from my advanced flying training at Valley (where he had been one of our QFIs) and we got on very well throughout our time together at Scampton. Officially I was Simon's deputy commandant Central Flying School although this role made no real demands. JACIG was headed up by a fellow group captain and TMTS was well run by an experienced flight lieutenant. I had a bunch of capable wing commanders on the station and a very able civilian contract manager for engineering support on the flying side. By this time, Scampton was a relatively small station with only about 700 personnel, evenly split between service and civilians. Good support from my immediate subordinates enabled me to focus on running the station and to fly which I did regularly.

The only shadow hanging over this almost idyllic inheritance was a rumour that the station was to be closed down. In fact, I had met the commander-in-chief, Air Marshal Sir Christopher Coville, at Scampton whilst doing some of my flying training there before assuming command and he had told me that, notwithstanding the strong possibility that the station would close down, a decision had been taken that I would still take over the station as planned. I was therefore well prepared for the fact that this would be a shortened tour which, unfortunately, it was.

In the mid-1990s, the job of station commander came with an impressive amount of support: a large six-bedroom house; a staff including a PA, a driver, a housekeeper and a batman; and a small budget for entertaining visitors to the station and local dignitaries. Having introduced myself to the officers in my first week and told them what I expected from them, and got to know my wing commanders, I quickly settled into a routine of keeping up with the various issues crossing my desk, routine inspections, disciplinary hearings, interaction with the local community and staying involved with the flying on the base, both with the Central Flying School and the Red Arrows. Whilst I was qualified as a CFS instructor on both the Tucano and the Bulldog, the Tucano squadron was normally fully manned whereas there was more scope to help out on the Bulldog squadron. This suited me well as the Bulldog, although an elementary training aircraft, was a demanding aircraft to fly well and an interesting change from many years of flying fast jets. I also did some examining whilst flying for the Bulldog squadron to grant the students their B2 (probationary) QFI qualification on graduation. Meanwhile, having got myself checked out on the Chipmunk, I was then current on all four aircraft types on the base. It was delightful to be flying the Chipmunk again, my first aircraft, which I had not flown since I left Barnwood.

In December, the Tucano squadron planned to take four aircraft to Wiesbaden in Germany on a staff training flight and asked if I would like to join them (essentially getting the station commander involved would ensure that the trip was given approval). We had a good flight over and an enjoyable night out. The next morning over breakfast in the hotel I was telling the other pilots about our Harrier battery issue

Author back flying Chipmunks, RAF Scampton, summer 1995.

in Madrid in January; we then went off to the air base to submit a flight plan and prepare the aircraft for the flight back to Scampton. When we arrived at the Tucanos we found that the battery switches on one of the aircraft had been left on and the battery was, of course, now dead. A call to Scampton and we got a message back very quickly to say that an RAF VC10 was being diverted into Wiesbaden with a new battery. Amazingly the aircraft arrived late morning and we were soon on our way back to base – an impressive bit of logistic support.

I flew the Hawk regularly and accompanied the Red Arrows on many occasions, both on training flights and on displays; I also flew their spare aircraft to some of their display venues. I remember well flying with them on their display at Dartmouth in 1995 and going down to the town after the show for the evening which was a great night out. I also flew out to Jersey for the Battle of Britain display in the September. A few of us were very fortunate to be invited over to Guernsey for a memorable hangar party. We were flown over in a twin-engined VIP turboprop and as we taxied out for take-off Squadron Leader John Rands (Red 1) jokingly asked the pilot where the drinks were. He pointed to a drawer and sure enough there was a selection of alcoholic drinks to keep us going on the ten-minute flight to Guernsey. The following morning we had a very early start to get us back to Jersey as the aircraft was required for another job. As we walked out to the plane I was chatting to the pilot and once he found out I was a pilot he said I could fly the aircraft on the return leg. Lynne Bostock (Simon's wife) was just behind me and she was more than a little alarmed telling Simon, "Chris can't fly us back, he's still drunk from last night!"

Life in the office was never dull and I was very fortunate to have an excellent OC Administrative Wing, Wing Commander Dave Gardner, whose commitment to the station and supporting the flying effort at Scampton was outstanding. Whatever cropped up, Dave was always there with good advice and an unstoppable humour. He was also the president of the officers' mess committee and resolutely refused to give up his appointment after his allotted time; amongst other considerations, the bar profits would have dropped noticeably if he had not been in the mess on a very regular basis.

Early on in my tour I had an informal visit to the station from the Air Officer Commanding, Air Vice-Marshal John May, an ex-Lightning pilot and a very affable officer to answer to. I was informed that the AOC would be at RAF Finningley that morning on an AOC's formal inspection and he would be flying into Scampton on a Jetstream arriving at 1330. Around 1230 I left the office for the residence so I could get some lunch before meeting the aircraft on the airfield. As I walked into the house the phone rang with my office informing me that the AOC was about to land. I jumped into my staff car – no driver or 2-star staff plates – and rushed out to the airfield where the Jetstream was just taxiing in. I met John May as he came off the aircraft, apologising for the lack of star plates or driver. He shrugged all this off and we drove the short distance down to CFS Headquarters to start the visit 45 minutes early, everyone having to scramble around to meet the revised timings for the visit. Despite the unexpectedly early start, the visit went well. Once I had seen the AOC off I phoned the station commander at Finningley, Group Captain Dave Wilby, to ask him what they were doing dumping the AOC on us an hour ahead of the agreed time? Dave apologised profusely and explained that the AOC had stayed with him the night before and they had been up all hours drinking whisky. Consequently, John had spent the formal visit in a somewhat subdued manner, sipping glasses of water, and when he was asked what he wanted to do on the Jetstream en route to Scampton, which was allocated for an hour's training, he just said, "go there and land" which took all of ten minutes.

One day my OC Station Services Flight came into the office to tell me that we had a problem with a very small storage site nearby that apparently I had responsibility for; leftover from World War II, inter alia, this location had been used for the storage of mustard gas. The current issue was that some travellers had decided to set up camp there with no knowledge of the history of the site. My officer had visited them and explained that they were illegally camped on MoD property. They had been very pleasant, and had even offered him a cup of tea (but having seen the state of their mugs he had politely declined) but showed no inclination to move on. They did leave eventually of their own volition and without anyone having to explain the true situation. I also discovered by chance that as station commander I had responsibility for RAF Binbrook, the former Lightning base in Lincolnshire. Interestingly, I did not receive any sort of brief from Support Command Headquarters at Innsworth before (or after) taking command of the station at which these responsibilities might have been covered. I doubt my predecessor knew anything about them either – he certainly never mentioned them to me before he left.

At Christmas, HQ decreed that all Support Command stations would work right up to lunchtime on the 23rd of December. This did not go down well on the station as we had been expecting to finish work a little earlier. To try and make this more palatable, I said that those who wished to could take leave and on the flying side we would do some staff continuation training which would take the form of two formations of Tucanos, one of which would visit the southern RAF stations and one the northern stations. It was a clear but cold day and I flew in the second formation which went up north. As we headed back towards Scampton we got a call from air traffic control to say that fog was forming on the airfield (it had not been forecast) and we needed to get back as soon as possible. When we arrived, we could see that the fog had formed below the airfield, at the bottom of the Lincoln Edge, and a very light breeze had pushed it up the hill and onto the south-west side of the airfield. The fog bank was now steadily making its way across the airfield. As we broke into the circuit, the ATC tower was swallowed up by the fog and we were now left with less than half the main runway to land on. Fortunately the threshold of runway 23 was still in the clear and we had no problems landing but rolled into the fog shortly after touchdown – a memorable end to the flying year! Having found our way back to dispersal in what turned out to be very thick fog, we finished work with lunchtime drinks at refresher squadron before starting the Christmas holiday.

As we moved into 1995 with confirmation of the closure of the station, plans had to be ramped up to make this happen. The Central Flying School was scheduled to move to Cranwell in the spring and the Red Arrows were to move there as well at the end of the season although they would continue to use the Scampton airspace for their training flights. Dave Gardner visited the office to discuss the non-public funds owned by the officers' mess – i.e. profit accrued by the mess which was effectively owned by the mess members. Any of this money remaining when the station closed would go to central funds and therefore mess members would not get any benefit from it. We decided that the best way to deal with this would be to hold a spring ball prior to Central Flying School leaving the station, without prejudice to the mess having a summer ball, as usual, in the summer. And this is what we did. We had two excellent mess balls that year which were very heavily subsidised and unsurprisingly very well attended.

During this period, CFS had a liaison visit from the Civil Aviation Authority Examining Unit. I asked OC Exam Wing if they could arrange for me to renew my civil flying instructor's rating during the visit. This was duly arranged with a venerable ex-CFS instructor and examiner called Dickie Snell who had been involved in bringing the Bulldog into service at RAF Little Rissington at the same time I had been there, although I had not met him at that time. The pre-flight brief went well and, before we took off, I explained that we had to be back on the ground by a given time as the Red Arrows would be carrying out a practice display at the airfield and we would not be able to land. Even so, we had plenty of time to carry out the test. However, once we got airborne, Dickie was enjoying flying the Bulldog again so much that he wanted to prolong the flight and asked if we could do that. If we were not to run out of fuel, the only way to do it was to land at Sturgate where Hugh Stark, my old squadron commander from Valley, was based, have coffee with him then continue the flight once the Reds had finished. We duly landed at Sturgate and got a coffee at which point Dickie asked Hugh if we could borrow a briefing room and some paper and pens so he could do some ground school with me as part of the civilian instructor's

RAF Scampton photograph taken prior to commencement of the station run down in spring 1995.

examination. This was not what I was expecting. Nevertheless, Dickie took me aside and proceeded to show me how little I knew about civil air law and propeller theory, all of which I had studied some years previously but had obviously now forgotten. We had an enjoyable flight back into Scampton where we taught each other how to fly an instrument landing system (ILS) and he then signed me up as a flying instructor. The next day I had the fully justified embarrassment of being told by a couple of the CFS examiners how Dickie Snell had delighted in coming into their office late yesterday afternoon to tell them how clueless their station commander was at ground school.

Just before Central Flying School left Scampton, being qualified on all four types that we flew on the station, I took the opportunity to fly all four aircraft in one day. Simon Bostock and I flew the Hawk before the 0800 Met Brief, we each flew a Bulldog to Cranwell as part of the move of CFS, we then brought the Chipmunk back to Scampton and flew a Tucano together in the afternoon. For good measure, I also flew three times with the Scampton Gliding Club that evening.

With the departure of CFS came the question of what to do with the two Chipmunks at Scampton as there was no room for them at Cranwell. One went to Coningsby, as they used the Chipmunk for tail-dragger training for the Battle of Britain display pilots, and I glibly said we could keep the other one at Scampton, knowing full well that I would be the only person qualified to fly it. After a couple of months of using the Chipmunk as my personal hack, I thought I had better come clean and seek authority from HQ to retain the aircraft. I had some trouble concocting a viable case for this but I did receive budget cover to keep the aircraft at Scampton until my departure in November. At this point I returned the last Chipmunk to storage at RAF Newton, where I used to do some of my air experience flying as a CCF cadet from Uppingham – full circle. Amongst other times I used the Chipmunk was visiting John May at Innsworth for my annual report, a somewhat formal interview. He was quite surprised when I entered his office in my flying suit but when I explained that I had flown myself down in the Chipmunk to Staverton, his only comment was "oh – that's good!". He wrote in my logbook summary for the year: 'A very creditable flying rate' although I suspect in the view of some other AOCs, this would not have been a positive remark on a station commander. I also flew myself to RAF Halton to take the salute for a recruit passing-out parade at RAF Henlow for Oli Delaney, an old friend from Cranwell and Greenwich, who was station commander.

At one of our regular dining-in nights, I was giving the after-dinner speech when I noticed that Warrant Officer George Brodie, the mess manager, had come up to Dave Gardner who was sitting next to me and was busy talking to him which was rather odd. As soon as I finished speaking and sat down, Dave told me that one of the CFS instructors had committed suicide at home. Did I want to cut the speeches short or should we continue until we had finished with the speeches? I said we would carry on and sort it out once we were out of the dining room. Getting to the bar, I found

Angie Sargeant, my volunteer acting station commander's wife, and asked her if she would accompany me to the house. We arrived at the house to find an understandably very distraught wife, who thankfully I knew already, the mother of two very young children. The station staff did a great job of supporting the family with the aftermath of this tragic event and sorting out the funeral arrangements.

With the rundown of the station, given the reducing numbers of personnel it did not make sense to keep all the mess facilities running. Dave Gardner and I discussed at length the options of combining the corporals with the sergeants, or the officers with the sergeants, and whether to centralise messing in the officers' mess or the sergeants' mess. In the end we decided to bring the sergeants into the officers' mess although this was a rather difficult concept for some of the army officers in JACIG to accept. The first test for how well this would work was a dining-in night. I elected to make this a voluntary event which, perhaps understandably because it was the first combined event, was not particularly well attended by either the officers or the SNCOs. Nevertheless, I took the view that if it was successful then the next one would be better attended and that is exactly what happened. We also had a 'Farewell to Scampton' dining-in night to which a number of dignitaries and ex-station commanders were invited. I asked Dave Gardner and his team to seek volunteers to host our guests for this event; we had no problem getting volunteers from the officers and the SNCOs and had an excellent evening. To celebrate the closure of the station, we also held an all-ranks hangar party preceded by a fly-in from the Battle of Britain Lancaster and a display by the Red Arrows.

One major issue we had to consider with the closure of Scampton was what to do about Nigger's grave[29]. The simple answer would be to dig up the grave and move the remains to a new location; but what if there were no remains and Nigger was not actually under the headstone? How long does it take for the remains of a buried dog to disappear without trace? Would ground penetrating radar help us in our quest? So many questions and no obvious answers. One morning Dave Gardner came into my office to tell me that he had the solution saying: "no greater love has an officer for his station commander that he lays down his dog for him!" Dave's answer was to sacrifice his dog and bury him before any excavations took place so at least there would be a dog to move. In the event, since it was evident that the MoD would be holding onto the site for some years to come, the answer was to leave the grave well alone. This issue did give rise to the only phone call I got from the AOC-in-C, Air Chief Marshal Sir Sandy Wilson, who contacted me on a Monday morning after the problem of Nigger's grave had featured in the *Sunday Express* the day before.

One day I got a phone call from Lynne Bostock to say that Simon had flu and could I step in and help host a barbecue the next night at their residence for Carolyn Grace? She was bringing her Spitfire to Scampton and Simon had arranged to give her a trip in a Hawk in exchange for a ride in her Spitfire and would I be happy to help out

With Carolyn Grace and the Grace Spitfire.

there as well? I didn't need asking twice so had the amazing experience of getting to fly in Carolyn's two-seat Mk IX Spitfire. Although the weather was not great, we did get to fly over the Lincolnshire Wolds and managed a victory roll before landing back into Scampton.

It had been decided that the Red Arrows would go on tour in South Africa at the end of the summer, routeing via Turkey where they would also carry out some displays. It was obvious that they would need a senior officer to accompany them and provide some 'top cover' and, with my dwindling responsibilities at Scampton, I thought I would be the ideal man to take on this role. My suggestion to Simon Bostock that he really couldn't afford to be away from UK for three weeks and, as his deputy, I should take on the job unsurprisingly fell on deaf ears. Shortly before the team departed, I had a phone call from Squadron Leader Tony Cunnane the station and Red Arrow's public relations officer. Tony asked me if I would like to fly down to Cape Town courtesy of British Airways (BA) and meet up with the team there. It transpired that the Red Arrows had a close liaison with the airline and BA had very kindly offered me a seat on one of their 747s so I could get down to South Africa for the weekend. The captain of the 747 turned out to be a good friend of my brother-in-law (another BA captain) and consequently I was very well looked after. We flew down to Cape

Town overnight Thursday/Friday, I stayed with the crew in their hotel out of town and went out for an excellent meal with the three pilots locally on the Friday night. On Saturday I joined up with the Reds and flew in the back of Red 8 (Spike Jepson) on display. We ran in for the display over Table Mountain, which was gridlocked with cars and spectators, the venue over the harbour was stunning and, on the flight back to the airport we overflew one of the townships, the size of which was truly staggering. After the show we went into town and met up with the three BA pilots for an evening on the waterfront (which was in stark contrast to the township we had overflown a few hours earlier). The next day, the BA captain and I visited the Stellenbosch wine area (dramatic and beautiful) and then crewed in for the flight back to UK that night. It was a very short but a most memorable visit.

With the impending closure of the station, I was asked to write a foreword to a special publication being produced by the *Lincolnshire Echo*.

SALUTE TO SCAMPTON

A *Lincolnshire Echo* Special Publication 1 April 1995
Foreword by the Station Commander Group Captain C. C. N. Burwell MBE

It is undoubtedly a great honour to be given command of what is, arguably, the most famous of all the Royal Air Force's stations. It is also a somewhat humbling experience.

Below: The Red Arrows fly over the Battle of Britain Flight Lancaster at RAF Scampton in July 1995.

Opposite: Last flight with the Red Arrows on 22 November 1995. It ended with the customary champagne and a drenching by the fire crews.

Naturally, we all recall the outstanding feat that 617 Squadron performed in May 1943. But the Royal Air Force station at Scampton is a testimony to much more human endeavour than this one act. From the days of the First World War, through the bomber era (1936–1982) to the present day, Scampton has been home to many generations of servicemen and local civilians, all of whom have served the RAF and the nation in war, and in peace, in their own way.

> Although I am a relative newcomer to Scampton, I have spent half my service career in Lincolnshire and I am well aware that the county and the RAF are synonymous. It is, therefore, with extreme sadness that we now face the closure of the station.
>
> I am privileged to be in command of RAF Scampton, not just for its proud history, but also for the dedicated and professional workforce that I have with me today; and in this respect I have no doubt that they are most worthy successors to their forebears.
>
> I hope you will enjoy this history of our station written by Peter Jacobs. It is a fitting tribute although, sadly, the timescale of its publication is wholly apposite.

As we approached the final rundown of the station and my planned departure from Scampton in late 1995, my posting officer contacted me to tell me that I was being posted to RAF College Cranwell as director of Initial Officer Training. This was a job that I thought would suit me well and I was pleased to get the news. However, he called me again a few weeks later to ask if I would be happy to attend the Royal College of Defence Studies (RCDS) course in London the following year instead. Once I had discussed this with Oli Delaney, who had previously completed the course, I could see that this would be a very interesting way to spend the coming year and a truly broadening experience. My posting was therefore changed to the 1996 RCDS Course starting in January. Leaving Scampton in November 1995, I was not to fly again until January 1999.

THE CO'S PARTING GRACE

by The Reverend Alistair Bissell RAF Scampton

Here we offer a simple grace
For all who find it hard to keep the pace
The time has come for some to go
We applaud their part in Scampton's show
To the CO Chris for all you've done
For unique contribution and gratitude won
Around the Station or flying his mean machine
There's little dirt on him
He's almost squeaky clean
So God grant to all a tasteful night
And in the morning may our hangover be slight.

CHAPTER 11

HIGHER STAFF TRAINING AND RAF LOGISTICS COMMAND 1996–99

The RCDS course was conducted at Seaford House in Belgravia and it had been designed to prepare selected officers for higher rank (2 stars and above). Running from January to December with around 80 students, half of them were from the UK (army, navy, RAF and civil servants) and the other half were from around the world, once again a mixture of military and civilians. So besides the immense breadth in the content of the course, there was the opportunity to get to know people from many different nations and gain an appreciation of their perspective on global issues as well as their regional concerns. This is a prestigious course and attracts many eminent speakers including academics, heads of industry, politicians, ambassadors and senior military officers.

Royal College of Defence Studies South East Asia tour embarking on a Pakistan army Puma helicopter, summer 1996.

On the Line of Control in Kashmir, September 1996.

The first term up to Easter was spent studying the United Kingdom as an example of a democratic state. This was followed by the regional tour where groups visited different areas of the UK for a week, they then presented their impressions and experiences back at Seaford House. I went to Northern Ireland which was a fascinating place to visit, especially in light of the Troubles. We met with politicians at Stormont and visited schools, a prison (holding both loyalist and republican prisoners together), industry, the army (who took us out on patrol in armoured personnel carriers) and met with local councillors and community leaders.

In the summer term we studied all the regions of the world culminating in the world tour where, once again, we split into groups to visit a region of the world, reporting back at RCDS afterwards. I was very fortunate to go to South East Asia – Pakistan, India and Nepal, three very contrasting countries. Similar to the regional tour, we were privileged to visit education, industry, the military, religion and politicians in all the countries visited. This tour provided a fascinating insight into the regional tensions between India and Pakistan, the Islamic State of Pakistan and the precarious position of Nepal both economically and geographically. Although the schedule was pretty busy, we did get a little time for some excellent sightseeing. Amongst many other very interesting visits, we got to see the Line of Control in Kashmir and had a most memorable trip up the Khyber Pass, through the Tribal Area (with an armed

The author with Captain Alan Phillips RN on the RCDS South East Asia tour.

military escort including an ambulance) to the border with Afghanistan, just after the Taliban had come to power. In the final term we looked at major world issues and the future. The year was a truly outstanding experience although I had to admit to myself that there were some people on the course who stood out as future very senior leaders and I certainly did not see myself in this category. Nevertheless, I was selected to attend the Higher Command Staff Course (HCSC) at Camberley in January–April the following year.

Similar to the RCDS, the HCSC was designed to prepare senior military officers for higher command in war fighting. Again I was rubbing shoulders with some of the brightest and the best and, whilst it was undoubtedly an excellent course, I regret I probably didn't bring much to the party. Nevertheless, the battlefield tour in France and Belgium at the end of the course was a poignant, interesting and enjoyable week away before returning for the final war game. Escaping from Camberley in April 1997, I moved on to my new appointment as Group Captain Plans at HQ Logistics Command based at RAF Brampton.

As I arrived into a headquarters focused on an area I had no background in, I was a little discomfited to find that my predecessor in the job had not even taken the trouble to prepare any sort of handover (I always planned and scripted a full week's handover whenever I left an appointment). He was moving to another job at Brampton and although I spent a week with him before he took up his new post, in that time I gleaned little about what my new job entailed. Nevertheless, once I was in post, my

At the Pakistan/Afghanistan border on the Khyber Pass, September 1996.

air commodore instructed me to set up a series of visits to Logistics Command units to get an understanding of the business which was time well spent. Another somewhat unsettling early event was an office call by a fellow group captain in the HQ, who asked me who I had upset to be working for my particular 1 and 2-star officers. They had reputations for not being the easiest of people to work for.

The scope of the Plans Department was broad, encompassing: preparing briefs for the commander-in-chief, estates management and rationalisation, IT policy, and public relations. Unfortunately, as had been the case at Barnwood, there was no real role for the group captain in this organisation either. Each of the discreet Plans areas were headed up by a wing commander (or civilian equivalent) and their output was fed up to the chief of staff (2-star) through me and the 1-star. The 1-star interfaced directly with the 2-star making the group captain post not much more than a letter box; in fact in some areas, the output went straight to the 1-star without my involvement. I was also bemused by the fact that the commander-in-chief had insisted on having a general duties (i.e. aircrew) officer filling the Group Captain Plans post since I never worked out what value my background was bringing to the job. (The broader view would be that I was learning about the wider air force to fit me for future appointments and there is no denying that in war-fighting logistics is a hugely important element.) I have to say that I did not find the work interesting and I did not enjoy it. Settled into the job, one evening just before I left work, my 2-star called into my office; this was a rare occurrence. He undoubtedly called to ask me about something (a social call was not in his nature), but I thought that while he was there I would take the opportunity to explain that I was not exactly fulfilled in the job and his assistance in moving me into something more interesting would be appreciated. A raise of the eyebrows was the only response I got. It was disappointing that he would not even enter into a discussion on the issue.

My two bits of light relief during this tour were flying the Bulldog (and later the Grob Tutor) with the Air Experience Flight at Wyton and going on a paragliding course. I had always been interested in paragliding but with the demands of work and a busy personal life, I hadn't managed to devote any time to this. One day in the office at Brampton, a Defence Council instruction crossed my desk which included details of a paragliding course being run in the Brecon Beacons in South Wales. I phoned up straightaway, only to find they were fully booked, but somehow managed to persuade them to put me on the course. My air commodore was very reluctant to let me have two weeks out of the office but eventually he said I could go, but would have to do some work whilst I was in Wales and, if required, I would have to be back in the office for the second week. I could not understand what he thought was so important about my job that it was so difficult to release me. Anyway, in the event I managed to stay for the full two weeks. The course was a great experience. First we had to learn how to 'fly the wing' in a flat field. This was quite a trick but, once I had the hang of it, the instructor told me to face into wind and start running; he then came round behind me and gave me a hard push at which point I rose into the air by some ten–15 feet – a great feeling – what an amazing wing! After that we went out onto various slopes getting bigger and bigger until we were launching off the lip of a 700-foot hill practising turns, spot landings, 'big ears' (when you partially collapse the wing tips to kill lift), ridge soaring and eventually landing back on top of the hill having made a height gain. By getting up very early in the mornings of the second week, we managed to get some great flying conditions with light winds ideal for beginners; by mid-late morning the wind would be too strong and we would be finished with flying for the day.

Back at work I now had to make a decision whether I stuck with the job in the hope of getting promoted to something more interesting and enjoyable or whether I elected to leave the RAF and do something else with the rest of my working life. I had no wish to move around staff jobs as a group captain for the next eight years just to get my full pension at the age of 55. My decision was largely influenced by whether I was a serious contender for promotion and, if so, what my job prospects at the next level were. Consequently I arranged a visit to the Personnel Management Agency at Innsworth for a career interview. There were only three jobs I was seriously interested in at the 1-star level (and by background I was a strong contender for all of them) but I knew that two of them had recently changed over. Shortly before going to Innsworth, I discovered that the only remaining job coming available in my timeframe (commandant Central Flying School) had been allocated to an officer without any background as a flying instructor. Therefore, my mind was largely decided before I went to the career interview. At the interview I was told that with five years in rank, not much over two years productive service as a group captain[30], I still needed one more good report to be a contender for promotion. It was a relatively straightforward decision to apply for PVR from the RAF although the lack of certainty about what I was going to do added a certain frisson to the situation. To give the RAF their due, I did get a phone call from the Air Secretary at Innsworth a short while after applying for PVR to ask me if I would reconsider my decision; he added that he had a job in mind for me. I did not press him on this and explained that my decision had been made and I was not going to reverse it.

I had been good friends with John Danning from our time at JSDC and, over the intervening years, we had got together for various walking trips with Paul Hopkins and Dave Harle. John had left the RAF a few years previously and was now manager of Cobham's base at Teesside. He had told me some time back that I should get in touch if I was leaving the RAF and was interested in working for Cobham. Considering that my expertise was in flying, and the management of flying operations, and I was keen to return to get back to the cockpit, speaking with JD about the possibility of working for Cobham was an obvious place to start. John was as good as his word and I was very quickly offered a position at Teesside as a direct entry captain on the Falcon 20 with the option (between JD and myself) for me to take over his position as base manager if he should move down to head office at Bournemouth as director of Flight Operations (DFO).

CHAPTER 12
THE TEESSIDE YEARS: COBHAM 1999–2011

I left the RAF in July 1999 and, having acquired my civilian instrument rating, commenced my training with Cobham. After two weeks at Bournemouth and three weeks in the Falcon 20 simulator in Paris (Le Bourget), I was ready to join the fleet at Teesside as a first officer, in training for a captain's position. The Cobham operation in support of the Ministry of Defence derived from the Falklands War (1982) when a number of Royal Navy ships were successfully attacked by Argentinian fighters using Exocet stand-off missiles and iron bombs. Following this hard lesson, the MoD contracted Cobham to provide target facilities training for the Royal Navy. The RN was supported by Cobham out of Bournemouth operating into the Channel and South West Approaches area. In the mid-90s, the contract was extended to provide operational training for the RAF and allied air forces. This element of the contract was supported by the company out of Teesside using six Falcon 20s. JD had set up the Teesside operation and I joined it four years after its inception.

We flew the Falcons with a crew of three: a captain (the majority being ex-military fast-jet pilots); a first officer (usually young pilots who had been hours building as flying instructors with a view to moving on to the airlines); and an electronic warfare officer (EWO), all of whom were ex-military and generally they were the mission specialists on board. The aircraft had been specially modified for this role with all the 'unnecessary' weight removed (e.g. soundproofing, passenger seats and toilets), hard points fitted to the wings to carry the electronic warfare (EW) pods, an EW control suite in the back for the EWO, and additional fuel tanks fitted into the cabin. In this fit, the Falcon's all-up weight had been increased to 30,000 lbs to enable the aircraft to carry four EW pods and enough fuel to complete missions of up to three hours. Our normal operating areas out of Teesside were the east coast ranges, which extended all the way up north beyond Scotland, although we did fly all over mainland UK, Europe in support of NATO, and occasionally further afield in support of friendly air forces in North America, the Middle East and the Far East.

Having taken over the missions that had been flown by the RAF Canberras, as time went on and the Teesside operation proved its competence, we flew a much wider range of missions either as single aircraft or in formations of two or more aircraft, sometimes up to six. The types of missions we flew, which could also include communications jamming and spoofing, were:

- Radar-jamming training for ground-based air defence radars.
- Radar-jamming training for fighter radars.
- Radar-jamming training for AWACS (the E-3 Sentry airborne early warning aircraft).
- Embedded in operational training packages as additional assets, providing radar screening for other aircraft in the package and, if required, acting as decoys.
- Towing an air-to-air gunnery target ('flag') for fighters to shoot at with ball rounds (as opposed to high-explosive rounds).
- Simulated AWACS providing check-in and control procedures for RAF assets.
- An air-to-air threat (particularly for slow movers e.g. C-130s); often co-ordinated with another Falcon 20 providing simulated AWACS (above).
- Practice intercepts (training for fighter controllers).

After two months of fairly intensive training in the right-hand seat at Teesside, I was appointed to a captaincy, a position I filled for the next 12 years. The task at Teesside provided interesting and varied flying and all of us spent between eight and 12 weeks away on detachment every year. The majority of detachments were in support of NATO air forces or major NATO exercises across Europe. We deployed as a crew for the duration of the detachment (on base we changed crews all the time), had a car for each crew and normally stayed in quite good hotels. We tended to rotate the driving duties between crew members on a daily basis but it was noticeable that some EWOs' driving skills were decidedly suspect, which might explain why they had not become pilots. Indeed on one detachment to Portugal, the first officer and I banned our EWO from driving the car after scaring the life out of us one evening on the way back to the hotel. When not on detachment, we operated on a weekly flying programme and even though this was liable to change, we were only required to come in to work when scheduled to fly. Generally there was no weekend flying and little night flying, both in stark contrast to flying for an airline; the lifestyle was quite relaxed with only three to four days' work each week when not on detachment.

In early 2001, JD told me that he was moving down to Bournemouth as DFO and asked if I wished to take on his job as head of Teesside (HOT). I was very happy to do this. The HOT role was wide ranging: I would carry on flying but would have responsibility for the delivery of Contract 020 to the RAF out of Teesside and on detachment. I would also be responsible for our flight operations, recruitment (aircrew and operations and admin staff), real estate and budgets. In addition, since JD was the DFO for the flight calibration unit at Teesside (Flight Precision Limited – FPL), he wanted me to become a flight calibration captain on the King Air B200. By this means, I could oversee their flight operations and keep him informed of any relevant issues on that side of the hangar. Consequently, I undertook a B200 conversion in late 2001, learnt the flight calibration role as a first officer and became a captain in 2002.

I really enjoyed flying the B200 and the flight calibration role – the use of a specially equipped aircraft and crew to measure the accuracy of ground-based electronic navigational and runway approach aids – for four main reasons. Firstly, unlike the Falcon role, flight calibration was totally different from any flying I had done previously; secondly, the aircraft was a sturdy, reliable and fun machine to fly; thirdly, we flew to varied and often very interesting locations; and finally, the crews were a much more eclectic and relaxed bunch than the Falcon crews. The various contracts that FPL held took us all over the UK, Ireland and a number of European countries. An example of the variety would be flying in the west of Ireland for a whole week (one of my favourite detachments) calibrating not just at Cork and Shannon but also taking in smaller airfields such as Galway, Kerry and Donegal. Then, another week, we would spend afternoons conducting calibrations at various Dutch airfields whilst doing nights at Amsterdam Schiphol. At Schiphol we would be given a one-hour slot on the calibration runway and you would be co-ordinated with the commercial traffic taking off and landing on the other runways; the Dutch air traffic control was first rate. Another part of the FPL operation which I enjoyed was being given a number of tasks for a day, or a few days, or a week, and having the autonomy to organise how best to complete them. Since good weather was required for many of our tasks, there was on occasions a need to change the plan (and night stops and hotels) as we went along. We would move first officers across from the Falcon to the King Air which would broaden their experience and give them the opportunity for a captaincy which they almost certainly would not get on the Falcon due to their lack of role experience. We would also give first officers who were doing well on the B200 the opportunity to transfer onto the Falcon with a view to moving back onto the King Air as a captain. In this way we created a development programme for those who wished to spend a few years with Cobham before then probably progressing to the airlines.

I remember calibrating at Brussels International for a few days. This was quite a relaxed detachment as we could stay in the same hotel rather than changing hotels every night as normally happened. One day, around midday, we had a call from Teesside to say that Dundee needed an emergency calibration of their ILS and we had to get there that afternoon so we could start work in the morning. Having submitted a flight plan to return to Teesside to collect the flight calibration documentation for Dundee, we rushed off to the hotel to clear our rooms and pay the bill before setting off back to the UK. I put the very inexperienced first officer into the left-hand seat

(LHS) for the transit back (we always flew the aircraft from the LHS with the non-handling pilot in the RHS). When we arrived back at Teesside, the crosswind was right on the aircraft limit. I asked the FO if he was happy to carry out the landing to which he replied no. Now whilst we were all qualified to fly the aircraft from the RHS, this was purely a qualification we completed every six months and we never normally flew, or landed, the aircraft from that seat. With the throttles positioned close to the pilot in the LHS, landing the aircraft from the other seat was not straightforward due to the operation of the throttles especially in tricky conditions such as a strong and gusty crosswind. It was not possible to change seats in the air so I was left with no option but to carry out the landing myself. It was a relief that the landing went well but I put myself in the LHS for the next leg up to Dundee, a night-time arrival into a small airfield I had never been to before after a long and challenging day.

On another occasion, I was en route to Madrid with a refuelling stop at a provincial airport in northern France. As was normal, after landing I went into the terminal building with the flight inspection crew, whilst the FO (Craig 'Huffy' Hough) refuelled the aircraft and I went off to check the latest weather forecast. Huffy was new to FPL but was an experienced ex-RAF pilot. Once he had caught up with us and had a coffee, we set off for security to get back to our King Air. It was at this point that we discovered that a full load of passengers for a B737 was waiting to go through the single-point security but we could not wait for them or we would miss our take-off slot. In my best schoolboy French, I explained our situation to the security staff who, in best Gallic tradition, despite seeing that we were an operating crew in uniform, told us to join the back of the queue. At this point, Huffy said, "Don't worry, follow me," and led us out of the terminal building and down the road to a security gate where he entered a code, opened the gate and we all walked through airside and out to the aircraft without any security check. Huffy explained that having refuelled the aircraft, he hadn't seen which way we had gone into the terminal so he had exited through the security gate. On the inside of the gate, the security code had been clearly displayed which he had remembered – and this was French airport security at its best post-9/11.

Calibrating at Sylt in the Frisian Islands one January, I had an FPL captain with me acting as first officer (Andy Thourgood) and the rear crew were from our German company. The rear crew were keen to start calibrating as early as possible the following day so they could finish the task that day and get dropped off at their Braunschweig base. Somewhat reluctantly I agreed that we would have an early breakfast and get a taxi to be at the airport for 0800. Arriving at the small terminal building in the pitch black, with the temperature well below zero, there was not a light to be seen in the building and our King Air was covered in a heavy coating of clear ice. We jumped back in the taxi and went back to the hotel for a second breakfast before we froze. Once the airfield opened at 0900, we returned and got a meteorological report that indicated that the temperature would not get above freezing that day. Since there were no de-icing facilities at Sylt to remove the ice from the aircraft, we were not going anywhere. We spent the early part of the morning drinking coffee in the airport cafe and then Andy remarked that there appeared to be an engineering facility in the ex-military hangars on the far side of the airfield. We jumped into a taxi and went across to see if they could help in any way. Their suggestion was that if we could get our aircraft across the airfield, we could put it in their hangar and, given time, the ice would probably melt even though the hangar was not heated. Breaking through the clear ice to get into the King Air was not easy, but we eventually got the aircraft started and taxied cautiously along the very icy taxiways. Shutting down outside the hangar, we put a towing arm onto the nosewheel, manhandled the aircraft into the hangar and shut the hangar door. Within a few minutes we could hear the ice starting to crack as it expanded and there were definite signs of the ice beginning to melt. We went back to the terminal building for some lunch and when we returned, the King Air was clear of ice sitting in a big pool of water in the hangar. We paid the engineer €100, took the aircraft outside, started up and went off to calibrate the ILS.

We had an unusual occurrence on return to Teesside from this detachment. When you fly you are required by law to have an alternate destination in case you cannot land at your intended airfield. On checking the weather for our return, we could see that Teesside was forecasting heavy snow showers, as were most other airfields in the north of England, but we were still legal to return with Leeming as our stated alternate. When we arrived back into UK airspace, we found that Teesside had just had a heavy fall of snow and, although the weather was now clear, the airfield was temporarily closed due to snow clearance work on the runway. We had plenty of fuel so were prepared to wait for the runway to be cleared but Teesside could not give a forecast of when this might be. While this was all being discussed, we then found that Leeming had gone red (weather below limits for landing) in a heavy snow shower. So we now couldn't land at our destination or our alternate. A check of northern England confirmed that heavy snow showers were now affecting all airfields and many were SNOCLO (i.e. snow closed). After about 30 minutes, the snow shower cleared Leeming and the runway remained open so we elected to land there while we had the chance to get on the ground somewhere. It was getting dark, very cold and unpleasant so I left the crew in the aircraft, went into the Visiting Aircraft Flight to tell them we didn't need any fuel and called Teesside to check on the state of the runway. After a while, Teesside managed to clear enough runway for us to land on so I went back out to the aircraft, we started up and transited back. As often happened in such unpleasant weather conditions, our excellent ground crew marshalled us to the hangar entrance, signalled for us to shut down and towed us into the hangar whilst we were still in the aircraft.

Cobham Teesside aircrew and operations staff in 2006.

I had an amusing incident when we were calibrating at Robin Hood Airport (originally RAF Finningley) one day. In between flights I had some time to kill so I was wandering around the airport terminal. I was browsing in a travel agent when the lady behind the counter, seeing me in uniform, asked: "Are you a pilot?" When I replied in the affirmative she said: "I don't know how you do it. Flying scares the hell out of me." "Well, I've been flying for over 40 years and only crashed once," I responded. "Did you survive?" she enquired. It took her a few moments as I burst out laughing before she realised what she had said.

From a holistic management perspective, FPL was never straightforward. This was not helped by the fact that we had a sister company based at Braunschweig in Germany that held contracts in northern Africa and the Middle East. The FPL MD, who had a limited background in aviation, was overseen and tightly controlled by the Cobham management at Bournemouth, none of whom knew much about flight calibration. As previously mentioned, their DFO was JD who was now based at Bournemouth and he had no direct contact with the operation. Despite this strange arrangement, when I started flying with them in 2001, FPL was doing good business both in terms of delivering the contracts to a high standard and producing a reasonable profit. However, as the years progressed, various issues arose which served to undo this successful operation, not least the loss of their very capable marketing manager who was not replaced for an extended period.

By mid-late 2006, I was making representations to JD that the delivery of the task was becoming seriously dysfunctional. Consequently, at the beginning of 2007, I was asked by the Cobham MD at Bournemouth to take on the role of operations director for FPL, maintaining my role as head of Teesside. He said this would be for six months; I was to dedicate 90 per cent of my time to sorting out FPL, he would not increase my salary but he would 'make it worth my while'. In the event, I held this post for well over two years until FPL acquired their own air operator's certificate, became autonomous and subsequently failed as a business, although that was still some years away. To give the Cobham MD his due, he did ensure that I received generous share issues and bonuses for my additional work with FPL.

My time as operations director for FPL was a very interesting period but, unfortunately, not always for the best of reasons. FPL had held a controlling interest in the German flight calibration company for some time but had bought them out and now owned 100 per cent. The German company was a single-aircraft operation and, besides sorting out the FPL operations at Teesside, I wished to understand their operation and see how we could harmonise the two companies. It was apparent from my first meetings with the Germans that they had no wish to join in any harmonisation or efficiency plan. On my initial visit to Braunschweig as operations director I had a four-hour meeting with their MD and chief pilot which covered a number of issues including their aircraft which I was aware was going to require a new main spar fitted in about a year's time. I was assured that this was not a major issue – about three months' work and not too expensive, they said. On returning to the UK, I spoke to JD and suggested we got the engineering director (ED) at Bournemouth to confirm their forecast (I was sceptical of what I had been told in Germany). Despite a number of reminders from me, it took some months before the ED came up with the answer: the work would take at least a year and the cost would be in the region of one million dollars. Well this really caused a ruckus since, in acquiring the German company, Cobham had inherited a bunch of contracts which, albeit quite lucrative, could be lost at any time and an aircraft that was now, very shortly, going to be grounded for about a year. The German company's offices were rented so had no value. Questions were asked about who had done the due diligence on the aircraft during the acquisition process. Although the management tried (quite incorrectly) to pin the blame on my opposite number at Bournemouth, as far as I am aware, this vital piece of work had never been carried out. FPL was then left with the issue of what to do about replacing the German aircraft. The only answer was to divert a B350 (a larger version of the B200 which had much greater fuel uplift giving longer range and time-on-task), which had been acquired for the Teesside operation, to Braunschweig. As one can imagine, this decision did not go down well with the crews at Teesside who had been looking forward to receiving this aircraft.

Now that Cobham owned the German operation, it was self-evident that, given the synergy, there was a huge logic in harmonising the two operations to improve efficiency. Strangely the MD at Teesside did not appear to see this as a desirable aim and, understandably perhaps, nor did the Germans as they could see the potential for the loss of some jobs. However, JD was in agreement with my view and I was asked to visit Braunschweig with Colin Pease, one of the managers from Teesside, to review what could be done towards this aim and to provide a report with recommendations. This was an interesting visit which flagged up a number of issues, particularly one relating to crew resource management problems (which we knew existed but could not prove). We produced what we thought was a well-considered report that did not try to be too revolutionary, taking a view that we could make a start on change and build from there. (I was of the view that ultimately the two operations should join to form one company; it did not make sense to have a single-aircraft operation in Germany operating autonomously with all the overhead that that entailed). Although JD appeared to agree with much of what was in the report, I never got any feedback from the FPL MD at Teesside who studiously chose to ignore it. In the remaining time I spent as FPL's operations director, nothing of note was done to improve efficiency between the two operations.

Another issue that arose whilst I was working for FPL was the attempt to start up a lightweight aircraft/low-cost calibration option for Teesside. This made a lot of sense; unfortunately, the way it was managed didn't. The concept was kept under very close wraps for a long time and we knew nothing about it at Teesside. JD eventually told me, in strict confidence, the basic outline which was to use another operator's aircraft with their flight crew, and for FPL to put one of our flight inspectors with a lightweight calibration suite that Cobham was buying into it to carry out the task. Asking my view, my only input at this stage was that I was surprised that we were going to give away our intellectual property rights in this niche business to another operator. Some weeks later, by which time Cobham was well committed to the project, we were given visibility of the full plan which gave rise to two major concerns amongst the operators at Teesside (including myself): firstly, the inclusion of a planning assumption that the aircraft would always be flown by a single pilot; and secondly, the suitability of the other operator.

Regarding the first issue, although CAA regulations did not stipulate that you had to have two pilots for calibration tasks, it was only safe (and common sense) to carry two pilots for runway approach tasks, i.e. the calibration of approach aids that required very precise flying ending up very close to the ground. To do this, the handling pilot had to fly extremely accurately on instruments which, realistically, precluded maintaining a lookout for aircraft and obstacles and degraded situational awareness. (We did fly single-pilot calibration tasks for area aids when it was not necessary to fly very accurately down to very low levels.) Single-pilot flying had been included in the project to increase the profit margin but either no-one had queried this at the concept stage or, if they had, it had been dismissed. Once we got sight of this, on behalf of the senior flying management, I raised our serious and justifiable concern on grounds of safety (risk of collision with other aircraft or obstacles on the approach). The other company was planning to fly the calibration tasks using a single pilot with a 'safety pilot' responsible for lookout. This was in contrast to the FPL precision calibration flights which were undertaken with two pilots qualified on type and in role. As far as we were concerned, this was the only truly safe way to ensure that the handling pilot was properly monitored on runway approaches and that situational awareness and a good lookout were maintained. Another concern was that calibration flying can be complex and demanding and requires a good understanding of the technicalities. FPL pilots learnt their trade as first officers over a protracted period and were only selected for captaincy once they had proved their ability and had gained a wide experience. This company was planning to put their pilots straight into flying as captains without any role experience. This became a rather heated exchange of views as following the professional advice of the operators at Teesside and introducing a qualified second pilot would impact heavily on the financial modelling of the project. In the end I was told in no uncertain terms by an email from the Cobham MD at Bournemouth to leave the issue alone; this effectively came down to shooting the whistle-blower and ignoring the safety case we had put forward, a decision with which I was very uncomfortable as a professional pilot and, by this time, an experienced calibrator.

To address the second issue – the suitability of the other operator – it was decided that the new chief pilot, a very affable but relatively inexperienced pilot and manager, would visit them to carry out an 'informal audit' of their operation. Since there were much better qualified people available to carry out this 'audit' (not least my opposite number at Bournemouth who was a very experienced auditor with an excellent knowledge of CAA flight operations legislation), this appeared to be a somewhat cynical, line-of-least-resistance attempt to get a buy-in from the FPL operators at the 11th hour. As far as I recall, no issues arose from the chief pilot's visit and the project, which was just about to start trials with the new calibration suite, was set to go ahead.

The week before the trials were due to start I was calibrating at Newquay. My flight inspector for the detachment was the lightweight-calibration project flight inspector and he was due to start the initial trial flying with the lightweight kit the following Monday. He confided in me that he was not happy about flying with them as he was not convinced they would have the same professional flying standards as FPL, not least he was concerned about them flying single pilot. There was little I could say apart from suggest he raise his concerns with the MD and DFO – I had already raised my objections and had been told to shut up. I do not know whether he did this or not but that weekend, before he went to join them, the other company carried out a series of calibration training flights. On the final flight, with two pilots in front and two company personnel in the back, the aircraft collided with a light aircraft; all five people in the two aircraft were killed. This was a tragic end to the project. I regret to say that at the time, and to this day, I felt that the concept was ill-conceived, largely by people who understood balance sheets but had no real grasp of the technical demands of the business. Those who did understand the technicalities were not consulted until too late and their input was then ignored. Sometime later, as a wholly internal project, FPL acquired a DA42 for lightweight calibration a few years before the demise of the company.

Somewhat frustrated by the poor management of FPL, and having held the head of Teesside job for almost ten years and looking for a new challenge, I applied for the job as FPL MD when it became available around 2010. Besides having a view that much more should be done to bring FPL and Braunschweig together, I felt that FPL would benefit from having an operator who understood the role of the business unit as MD (FPL had their own finance manager and there were two finance directors at Bournemouth overseeing their finances). In the event, the job went to another 'businessman'. This was not a huge surprise to me and, in the event, it was a blessing that I did not get the job; the new incumbent did not last long as FPL was now beginning to struggle financially. Sometime after I left Cobham, I was told that the B350 had been acquired on a lease, the terms of which were such that FPL was going to find it very difficult to ever make a profit[31]. I suspect this was one of the main reasons FPL ultimately failed as a business.

In comparison to FPL, the management of the Falcon operation at Teesside was much more straightforward given that the company had been delivering the business to one main customer, the MoD (RN and RAF), for a number of years. With a largely ex-military management on the flying side and the majority of crews being ex-military, the competence of crews was rarely in doubt and the front-line customer was invariably very pleased with our support. At head office, the finance director was well versed in the RN side of the contract and had been intimately involved in setting up the RAF extension and JD, the director of Flight Operations, had a good knowledge of both the RN and RAF elements of the contract. The main problems I faced at Teesside were more parochial: a division between the two operations, north and south (similar to the split between the UK and RAFG Harrier Forces); and personnel issues at Teesside.

Once I had taken over at Teesside and had access to the information, I quickly realised that the crews at Teesside were working more each year by a fair margin than the Bournemouth crews. Indeed, some captains down south were flying well under 50 per cent compared to my crews which caused some resentment, especially since they were earning the same money or, in some cases, more. In addition, Teesside crews were all spending more time away on detachment each year in comparison with their opposite numbers. On top of this, Bournemouth personnel had an excellent canteen for lunch and the use of a Cobham Flying Club which owned a Cessna 150. Both these perks were provided at a very favourable rate. At Teesside, we had no canteen and no subsidised light aircraft flying and these two issues were ongoing points of contention for many of my people up north. Realistically, given the limited number of personnel – about 120 – we were never going to get a canteen at Teesside (even though one had been included in the original plan for the hangar upgrade). However, I did manage to improve the quality of the packed lunches provided to crews flying over the lunchtime period. With respect to the flying club issue, having confirmed the level of interest in light aircraft flying at Teesside, I put a case forward to head office to provide some funding to support flying at one of the flying clubs at Teesside. This was generously approved and a sizeable sum of money was made available for my personnel to give them some equality with their fellow workers at Bournemouth. Sad to say, and an unfortunate reflection on human nature, once this had all been agreed at some considerable effort, the actual level of interest largely disappeared and there was a very limited take-up on this scheme.

Another north/south divide that raised its head was the conduct of the Joint Maritime Courses (JMC) exercises that took place twice a year in Scotland usually detached to Lossiemouth or Kinloss. These were predominantly a naval exercise so the Bournemouth crews had the lead and Teesside operated in a supporting role. I was nominated as the Teesside detachment commander for the second JMC I went on and afterwards I wrote a very short report on Teesside's involvement, recommending a few areas where I believed things could be improved. I thought that with my relatively recent front-line experience and a fresh pair of eyes, I might have something to add to help improve our involvement. I was somewhat taken aback when JD told me that the DFO at Bournemouth (JD had not yet taken his place) had taken great exception to my writing a report pointing out where we could improve. The DFO had set up the Contract 020 RN-support flying when Cobham acquired the contract back in the 1980s. He appeared to be of the view that nothing had changed in the intervening years and his original modus operandi could not be improved upon.

A few years later, I came across the same intransigence with one of the Bournemouth captains who I knew well from the RAF. The radio communications on the mission with military exercise traffic had been poorly planned and some of it had been quite unprofessional. When I made this point at the debrief, this particular captain would not entertain my view and got quite upset that I should be suggesting that there might be a way of improving on 'what we have always done'. Subsequently I raised the issue after the detachment and a meeting was arranged to revise and agree a new communications strategy that worked much better. It was sad to see such resistance to change. Unlike the military where everyone moved every two to three years and things changed, usually for the better, it was too easy for some people at Cobham to settle into a job and to keep turning the handle, doing the same thing every year and being resistant to change. I have to say that due to the dynamic nature of our operations with the RAF at Teesside, which was always looking for us to develop into new areas, this tended not to be a problem on our side of the contract.

Personnel management issues caused problems for much of my time running the Teesside Falcon operation. The relatively relaxed demands made upon the crews should have been a godsend. However, there was a reluctance from crew members to provide flexibility to meet programming changes during the week (we did not operate a standby system for a number of reasons). This soon came to a head and I called all the Falcon crews together to explain that if they were not going to provide the flexibility to meet programme changes, the company would consider implementing a standby system. This caused something of a furore, even though this was quite within the remit of the company and the operation if required. In response, a number of the crews decided to join the British Airline Pilots Association Union (BALPA) and invite them to champion their cause. In the event, this changed nothing and we (the management) soon got the unofficial word from the BALPA representative that our crews had no cause for complaint compared with what their airline members had to endure. There were a couple of interesting follow-ups to this. Firstly, one of the ringleaders and main agitators of the BALPA issue at Teesside became an internal BALPA representative but gave up his union membership after a few years as it cost too much (and had obviously not made any difference). Secondly, towards the end of my tenure at Teesside, I got a phone call from the BALPA representative with whom I got on very well. He told me that he had been at Teesside the previous day and, having time to kill, thought he would come across and have a coffee with me. He then thought he better not as my crews would think he was consorting with the enemy. Unfortunately he left his Cobham role shortly after this time (and he was certainly of the view that the Cobham crews were onto a very good deal and had no cause for complaint).

I also failed to curry favour with the Falcon crews in two other issues: proposing a readjustment of pay for senior first officers (SFO); and holding one of my captains to account to meet his contractual obligations to detach to Turkey. Over the years, with annual percentage increases, SFO pay had risen to a very high level which encouraged them to stay at Cobham with very limited prospect of making captaincy. (Cobham was always well placed to recruit from the military, acquiring capable direct entry captains with lots of relevant role experience.) The 'stockpiling' of SFOs added to the company overhead and was quite unnecessary. My proposal was based on a pay freeze for SFOs rather than a cut in pay but was never adopted, probably because it was opposed by my opposite number at Bournemouth who had built her career at Cobham and had been fortunate to make the transition to captain. Although I had made the proposal 'management in confidence', the proposal was leaked out and put me in bad odour with the first officers, especially the SFOs.

The refusal of one of my captains to take a detachment to eastern Turkey due to his perceived view of the security threat caused another problem. Since he was refusing to meet his contractual obligations, having consulted the HR manager and DFO, he was directed to meet with me to explain his position. Since he was due to leave Cobham in a couple of months' time, he requested to leave the company early and we happily dispensed with his services. Given the short-notice change to the crewing for this detachment, I took his place and had a memorable two weeks in Diyarbakir in the east of Turkey. At the time I was aware that my hard-headed approach to these matters would not endear me to the crews; as always, I made a point of doing the right thing and not the popular thing. I was unapologetic in supporting the best interests of the company and I certainly opposed JD behind closed doors on the few occasions when I considered his approach was wrong, although the crews would not have been aware of this.

An issue that was new to me at Cobham was flying with female pilots. Whilst women had joined the RAF as pilots before I left the service in 1999, their numbers were very limited and I had not had any professional dealings with them during my military career. At Cobham, they were also few and far between. Prior to taking command of RAF Scampton in 1994, I had been on a visit to the directorate of RAF Recruiting at Cranwell and women pilots had been a very topical issue at this time. I particularly recall one graph which illustrated that the female pilot aptitude curve was clearly displaced to the left (lower scores) than for males. This suggested that the RAF should be able to recruit some very capable female candidates for pilot training but that their numbers would be limited in comparison to males. I believe that experience in the military over the years has borne out this forecast since some female RAF pilots have gone on to fly the Tornado, the Harrier, the Typhoon and probably, by now, the F-35 Lightning. As far as female pilots are concerned in the commercial world, many of them will have had little or no aptitude testing before commencing their flying training (this is the same for men and for women). My own, very limited exposure to female pilots at Cobham showed that, like their male counterparts, their training and testing under the UK CAA fitted them perfectly adequately for their duties as first officers. However, I never saw the aptitude and resilience that you would expect in military pilots but which we did see in some of our male first officers (indeed, a few of our Teesside FOs probably had the aptitude and could have made it through military training and on to fast jets if they had tried joining the RAF). Given the limitations of our own selection process for first officers (of which more shortly), this was not a surprise to me.

Having female pilots did, perhaps unsurprisingly, produce one management headache for me at Teesside: pregnancy. Since pregnant pilots are not permitted to fly in months one to three and seven to nine and have the option not to fly in months four to six, followed by up to 12 months maternity leave, a female pilot may be absent from the cockpit for around 20 months. After this absence, they have the right to return to their job. Also, bear in mind that as the manager, I only became aware of this absence when the pilot involved came into my office and told me that she was effectively grounded with immediate effect and was electing not to fly in months four to six. With only eight first officers on the Falcon, given the time to recruit and train a replacement, we decided it would not make sense to replace her, especially as we would have to offer her the job back which could mean making her replacement redundant and wasting all the training. We managed her absence by shifting her flying onto the other first officers, flying captains as first officers and loaning first officers from Bournemouth when necessary. In the end, this one female pilot resulted in our Falcon operation being short of one of just eight first officers, on and off, for some four years. Whilst I totally accept the concept of equal rights, under this legislation, such a burden falls very heavily on a relatively small-scale flying operation.

I was also responsible for the recruitment of various staff at Teesside including the aircrew. The recruitment of direct entry captains and EWOs from the military was pretty straightforward as they were known quantities[32]. Whereas the selection of first officers provided more of a dilemma. Our first officers (FOs) were generally recruited from young pilots who were working as flying instructors at clubs in the UK and were interested in a different (exciting and challenging) flying experience before moving on to a career in the airlines. I was aware that up until now the selection of FOs had been done purely on the basis of a 30–40-minute interview. I felt that we should put more effort into this area and attended a recruitment module on my MSc course at City University to inform this process. In the event, I came up with a one-day selection format. Since the aspiring FO candidates had little understanding of the nature of our flying (either on the Falcon or on the King Air) we started the day with a 30-minute presentation on our flying operations; this was followed by an HR brief on terms and conditions. Through the rest of the day, the candidates were given: an interview; a flight in the most suitable aircraft for them (if it was possible); time in the crew room to meet and chat with our aircrew; and lunch (an aircrew 'nose bag'). At the end of the day, we took inputs from everyone who had been involved with the candidates, from the admin staff who had met them and talked with them on their arrival, to the crew they had flown with, the people they had met in the crew room and the interviewers. Such a process could not identify a great pilot, but that is not what we were looking for. By virtue of the licence they held (frozen airline transport pilot licence – fATPL), every candidate was, in theory, capable of meeting the CAA requirements for an FO on the Falcon or the King Air. What we were looking for was someone who would embrace the challenging flying we had to offer and who would work well as part of the team. Bearing in mind our regular detachments, and that each individual would have to work and socialise harmoniously with two other

Author and First Officer Ciara McGurk with a Falcon 20 about to depart RAF Akrotiri for Sharm El-Sheik (Egypt) and Bahrain on 26 September 2001. This was two weeks after 9/11.

crew members for one or two weeks or more at a time when away from Teesside, their ability to integrate was very important. I can make no claim that this process produced a truly radical change to our FO selection standard but we were generally well satisfied with those that we did recruit.

Despite the various management challenges, piloting the Falcon was a great experience and I certainly enjoyed the flying I did at Teesside and the many varied detachments. We had a rather bizarre experience in September 2001 when we were deploying to Oman shortly after the Al Qaeda attack on the Twin Towers and the whole world was in a heightened state of security. Having landed at Athens for a night stop, the detachment commander and I were in the handling agents' office when we saw our aircraft being shown on the evening news on television. We asked the agent what was being said and he informed us that there was speculation that we were deploying through Greece with live missiles on board. (Few people appreciated that the 'missiles' under the wings were EW jamming pods.) Some frantic discussions then ensued between ourselves, head office at Bournemouth and the RAF, and it was quickly decided that instead of continuing to Oman we would divert to RAF Akrotiri in Cyprus before the Greeks decided to impound the aircraft. We spent the next week in the Hilton Hotel in Nicosia, flying the occasional mission in support of the RN in the Mediterranean, before it was decided that it was wise for us to continue to Muscat via Saudi Arabia and Bahrain.

I had another experience of 'risky shift' when flying the Falcon at Teesside. I was scheduled to fly a line proficiency check on a low-level mission with some C-130s starting near Newcastle; the Falcon chief pilot would be the check pilot and would also be acting as my first officer. The weather forecast did not look suitable for the planned mission but we were aware that the C-130s were airborne so there was some pressure on us to make the rendezvous and try and deliver the planned training. As I expected, once airborne the weather was poor and I suggested to the check pilot that it was not suitable and we should abort the mission. The check pilot, who was an ex-military F-4 pilot of some repute and was much more experienced than me on the Falcon, said that we should press on. Consequently, we flew on in some very bad weather with neither of us taking the decision to pull out as we both thought the other was doing the decision making. In the end the Falcon chief pilot decided he had had enough and we did pull up but both of us should have made that decision a lot earlier.

One mission I remember well out of Teesside was a routine exercise in one of the danger areas off the east coast of the UK operating as one of a pair of Falcons against a pair of Tornado F3s. All was going well and we were halfway through an engagement when we found we had lost both UHF radios providing our communication (and safety) with our other Falcon, the Tornados and the RAF fighter controller. This was most unusual to say the least and we exited the fight, rocking our wings to indicate that we had a problem. We quickly ascertained that we were unable to speak with anyone on any of the five radios we had onboard our aircraft. We selected our IFF transponder to 7600 (lost radio) and left the danger area heading back towards the east coast then routeing north towards Teesside. The weather at Teesside was poor and would necessitate an instrument recovery – but without our being able to speak to Teesside. Of course, there are procedures for such an eventuality which we started to follow, not for a moment in the past thinking that we would ever need to do this. Fortunately, all our radios started to work again before we got back to base and we recovered the aircraft safely without further incident. In examining the aircraft, our engineers discovered that an intermittent short circuit on the EWO's radio press-to-transmit foot pedal had been cutting out all the radios on the aircraft.

After the awful events of 9/11, there was renewed concern over the possible hijacking of airliners by terrorists; consequently, Cobham was tasked with helping to train the RAF fighters for such an eventuality. In due course, this was extended to include a no-notice exercise to test the whole of the air-defence system of the UK. I elected to undertake the first of these missions myself, which was launched out of Glasgow under a spurious flight plan to land in Iceland. We took an RAF fighter controller in the aircraft with us as an observer and to assess the operation from our side. Before leaving the Scottish Flight Information Region we declared a hijack situation on the radio and set the transponder to the hijack code. After acknowledging our radio call, a long pause followed until we were asked to turn round and head south which we did. Impressively quickly, two Tornados (Q1 and 2) out of Leuchars intercepted us, followed in due course by Q3 and 4. Whilst our expected route was to be south into Wales, the air defence organisation elected to head us east, directly across the main north-south airways structure. Although we were not aware of it, as we were on a discreet radio frequency, this resulted in multiple airliners having to change track and/or height to deconflict with our formation of a Falcon 20 and four Tornados. Nevertheless, all went well and we eventually recovered back into Teesside with the job done.

Over the 12½ years I was with Cobham I detached widely across Europe: France, Germany, Belgium, Italy, Portugal, Spain, Poland, the Netherlands, Norway, Denmark, Hungary, Greece and Turkey and to Oman. One year whilst I was in Germany I was asked if I could take a Falcon into a military base in Poland as a proving flight for a detachment the following year – the first time Cobham would have flown into Poland. Having researched it, the only IFR (i.e. bad weather) approach I could get hold of for the base was a faxed copy of a Soviet NDB (non-directional beacon) approach in Russian which was, of course, totally indecipherable to us. Consequently I elected not to make the proving flight. The following year, the Polish detachment appeared on our annual task and I elected to take this detachment, now being assured that the base had a ground-controlled approach (GCA). On arrival the weather was beautiful but I elected to carry out a GCA to check out the accuracy of the approach and the controller's English. This was revealing: instead of the continuous stream of instructions which we should have received, the controller spoke to us a maximum of three times on the final approach and his English was so bad as to be almost unintelligible! Fortunately the weather was good for the week we were there. We went back the following year and the weather was not so kind; we were very grateful to have the excellent Proline 4 avionics in the Falcon to back up the GCAs we received.

This second Poland detachment was also memorable for a barbecue in the woods we were invited to by the Polish fighter (Fitter) squadron on our final night at the end of the exercise. I had a very interesting discussion with the squadron commander who had been trained under the Soviet system and had done much of his officer training and flying training in Russia. I asked him how he found things now that the Polish air force was operating under NATO. He said he still found it strange and, in some ways, would be happier operating under a system he knew better, although he did acknowledge that his more junior officers didn't have the same problem and felt more aligned to NATO. A rather startling moment came when we were discussing our visit to their SAM-6 (a Soviet surface-to-air missile) battery that afternoon. They pointed out one of their pilots and said he had been shot down by their SAM-6 battery on the same exercise last year by mistake![33]

Arriving at Muscat in Oman in September 2001, I had a rather alarming incident. On selecting the undercarriage down, there was an almighty bang which shook the whole airframe although, somewhat surprisingly, we found we had three greens indicating that all three wheels were down and locked. I had never felt such a violent jolt through an airframe before and thought there must have been a major problem with the undercarriage; however, the three greens suggested that the gear had sequenced normally. Tony Morris was about to land in another Falcon so he came to join us in formation and carried out a visual inspection of our aircraft which revealed nothing untoward. Our subsequent landing was uneventful and all looked OK when we got out of the aircraft as well – a bit of a mystery. That evening the engineers put the aircraft on jacks and found that the hydraulic nose jack had failed. Consequently, when we selected the gear down, the nose leg had lowered under gravity in well under one second, hitting the stops with considerable force, instead of in a controlled manner in five or more seconds.

Probably the most enjoyable Falcon detachment I went on was when I returned to Oman in 2011. Most years we sent two Falcons to Oman to support the RAF and the Royal Omani Air Force carrying out a major air exercise. Since this was one of the best detachments, I had made a point of not taking advantage of my position as head of the Teesside operation to go on this exercise. By 2011, most of my crews had been to Oman at least once and since I was retiring later that year, I could not be criticised for taking one of the captain's slots. Roger Sunderland, an ex-RAF Buccaneer and Tornado pilot and a sensible, level-headed personality, was detachment commander and I took the second aircraft. Unlike an airliner, due to its limited range, a Falcon takes six legs and two days to get to Oman. Day one we routed through Germany, Italy and on to Crete for the night; next day we continued through Jordan and Saudi Arabia and on to Oman (ROAF Salalah). All went well until we arrived at King Khalid International (Riyadh) where Roger's company bank card was blocked when he tried paying for the fuel. At this point we were beginning to run out of crew duty time to get to Salalah that evening. We were unable to unblock the card as the accountants at Bournemouth had not informed us what the answers were to the security questions. Rather than risk the same problem with my company card, I used my personal Mastercard to pay for the fuel and handling charges for both aircraft. We eventually arrived in Oman just in time before our crew duty ran out and pretty tired after a demanding six sectors over two days.

Having had a couple of days off in our excellent hotel in town, we commenced exercise flying which went on for two weeks but with the weekend off – very civilized. The flying over the desert was varied and interesting with the opportunity for a bit of low-level sightseeing on the way back into Salalah at the end of a number of the trips. The only untoward event during the flying was when an RAF Tornado decided to come and join in close formation on Roger's right wing whilst I was flying in fighting wing on the same side. This effectively put the Tornado between the two Falcons, without any sort of brief or agreement, and with the Tornado unsighted on my aircraft. It was potentially risky enough doing some of the flying we did on the Falcon, without the option to eject from the Falcon if anything went wrong, without such bizarre and potentially dangerous flying from a combat-ready RAF fast-jet crew. I couldn't believe my eyes when the Tornado did this and I quickly told him to get out of our formation; this was followed by a full and pointed debrief on the phone post-flight.

All too soon it was time to pack up, leave our lovely hotel and head back to the UK. In preparing to leave, we had belatedly discovered that we were required to carry HF radio to cross the 'empty quarter' flying from Salalah back to King Khalid. The reason for this is the limited VHF radio coverage over this huge expanse of empty airspace. Whilst both aircraft had HF fitted, both radios were noted in the tech log as being unusable (they were not normally used anyway). I attempted to resolve this dilemma with ATC but this came to nought, so we set off back hoping that VHF coverage on our route would suffice. In the event it all worked out OK and the fact that we had inter-plane communication meant that if anything had gone wrong we would have been able to communicate the problem via the other Falcon.

This time we got through King Khalid OK apart from another handling agent trying to steal our business as our booked agent was not around when we arrived – he had gone off into town, but quickly reappeared when we called him. Getting out of Queen Alia International in Jordan was another matter. Due to the proximity of the border with Israel, ATC cannot release you for take-off until you have been given a clearance into Israeli airspace. Roger taxied out first but when he got to the holding point, they said he did not have a clearance so he was told to taxi down the runway and taxi round again whilst ATC tried to get his clearance. When our turn came, we were cleared for take-off even though we could see an airliner approaching to land. With his landing lights on it was not easy to assess his range but he looked reasonably close. Consequently, as I started moving forwards, I got my FO to query the take-off clearance and ATC came back and told us to expedite (i.e. hurry up). There was also some confusion with callsigns as Roger's callsign was similar to ours and the controller was having some difficulty with that. As I started turning onto the runway, I could now see clearly that the airliner was very close and we were placing ourselves directly in his landing path. As a result, I carried out an immediate 180° turn and cleared the runway and the airliner landed just as we cleared. Shortly after this we were given a further take-off clearance. Once airborne we were informed that we did not have a clearance into Israeli airspace and would have to return to the beacon at Queen Alia awaiting a clearance. I informed ATC that I would be filing an occurrence report due to the incident on take-off. I filed the report immediately on return to the UK and was informed by the UK CAA that they would inform me once they had a response from Jordan. After many weeks, I was told that the controller had been under training and she had been retrained in light of this incident. Hardly the correct response as this should never have happened with a 'screen controller' in place to supervise her. Very fortunately, the incident had occurred during the day, in good visibility, when it was actually very quiet for an international airport. The outcome could have been very different if it had happened at night. The final day, flying from Crete back to Teesside in European airspace with two stops en route was absolutely fine.

There is a view that transiting aircraft from the UK is fine until you get to the southern European countries (Spain, Italy, Turkey and Greece) and certainly beyond the eastern end of the Mediterranean and then it can start to get very difficult – this was certainly my experience. On a detachment to a military base in Greece, the ATC was truly abysmal. They were certainly challenged by the English language (the international language of the air, whatever the French may think) and if they did not understand your transmission, they would just ignore you. This was particularly worrying when you were trying to recover to the airfield on minimum fuel. On the other hand, I should say that the civil ATC around the Athens Zone, who we used a few times, was very professional and helpful. On another detachment, arriving into Istanbul, we picked up the active runway as usual on the ATIS[34] frequency and briefed for this approach. Shortly after checking in with the approach controller, we were allocated another runway for our arrival. No sooner had we briefed the new approach than we were allocated yet another runway (another rebrief!) and given radar vectors onto a 20-mile straight-in approach. At about 15 miles we were told to reduce speed due to other traffic landing across our approach path. By eight miles we were told to reduce to minimum speed. Although the visibility was poor, we had the other traffic in sight by about four miles and I could see that the other aircraft's runway was about 90 degrees off our runway and their touchdown point would be close to where we should be landing. This posed a definite risk of wake turbulence (and potential loss of control of our aircraft close to the ground), especially given our relatively small size compared with the airliner. Having confirmed my view with the first officer I elected to carry out a go-around which, worryingly, came as a great surprise to the controller.

For some years I had had my sights set on being retired by the age of 60, my thinking being that I wanted to retire whilst I was still fit enough to do some fairly challenging walking and cycling expeditions. In the event, I arranged with Cobham that I would stand down from my management duties in May 2011 and they kindly agreed that I could stay on for six months as a part-time (50 per cent) captain on the Falcon leading into full retirement in November, two months after my 60th birthday. This was a great way to finish my active flying career and I carried out my last flight on 28 October, finishing with a total of 8,423 hours flying.

CHAPTER 13
RETIREMENT 2011–14

I had been looking forward to retirement, having the time to do things I wanted to do, and the next two-and-a-half years did not disappoint. Obviously there was a limit to how many expeditions I could go on each year and I had planned to take on some volunteering work to do something useful in life. Shortly before I retired, my wife, Lynne, asked me if I would take on the job of finance manager of her PR company; this would just mean one day a week in the office and a bit of time working from home during the week if required. This was unpaid work but provided me with a useful and interesting focus as I moved into retirement. I had already started volunteering with the National Trust, helping to look after the Hudswell Woods across the River Swale from where we lived in Richmond, and I continued this through until 2014. In 2012 I also began some volunteer work with the Soldiers, Sailors and Airmen's Association (SSAFA) assessing claimants for charitable donations. To train for this I was required to attend a short course in Edinburgh. It was an interesting three days but I, amongst others, was surprised to find that we could only report fact when putting forward a case, we were not permitted to make any observations on what we saw or thought about the claim being requested. Whilst this was a worthy cause, after my training under supervision around the Catterick area, it became apparent to me that the money was not always being allocated very fairly. Coupled with a somewhat high-handed letter from the chairman of SSAFA at Christmas 2013, criticising volunteers for getting 'above their station', I was not unhappy to give up my role the following year.

My main focus over this period was, as I have said, to carry out a number of walking and cycling trips now that I was fully retired and over three years I carried out the following 'expeditions', all but the solo Kintyre and Cowal trip, with friends from a small group known as KOS[35]:

July 2011	Tour du Mont Blanc (Alps walk)
June 2012	Scottish Inner Isles Distillery Tour (bike)
September 2012	GR5 Stage 1: Lac Leman–Les Houches (Alps walk)
May 2013	Kintyre Peninsular Tour (bike)
June 2013	GR5 Stage 2: Les Houches–Modane (Alps walk)
October 2013	Kintyre and Cowal Tour (solo bike)

May 2014	Danish Islands Tour (bike)
June 2014	GR5 Stage 3: Modane–Larche (Alps walk)

I was asked by a number of people whether I was going to fly in retirement and, if not, whether I would miss it; the simple answer to both questions was no. When I finished flying in 2011, it was over 44 years since I had first taken to the air solo (in a glider at Swanton Morley), 42 years since my Chipmunk solo at Perth and 40 years since my Jet Provost solo at Barkston Heath. Over the intervening years, I had done a lot of very interesting flying and I had been in and out of the cockpit a number of times on ground tours and I was aware that I did not need to fly. I had gained a lot of satisfaction from the flying I had done but, if I was sufficiently interested in and challenged by what I was doing out of the cockpit, flying was something I could live without. When I was on ground tours, I had flown Chipmunks, Bulldogs and Grob Tutors with Air Experience Flights as a way of staying in touch with flying and as a welcome break from staff work but not because it was a necessity. My time at Scampton on the Tucano and the Bulldog was usually 'productive' (instructing or examining) and helped me retain credibility with the flying units; piloting the Hawk and back-seating with the Red Arrows enabled me to stay in touch with their operation. But after the demands of the RCDS and HCSC, once I was established at HQ Logistics Command, unhappy in a job I could not relate to, my mind refocused on flying and that had pointed me towards the next move in my career – to Cobham. But did I now feel the need to keep flying as I moved on from Cobham and into retirement? Not at all.

Then there is the reason why one flies. Early flying training is, of necessity, focused on the basics and it is essential to get these right. Of course, much of the Cranwell course fell into this category but, even from the outset, our training introduced the concept of 'using' the aircraft, aerobatics on all the early flights being a classic example. This developed the mindset of discovering the full potential of the aircraft and the ability, and the confidence, to operate it to its maximum performance. Formation flying and low-level navigation are other examples of using the basic skills to achieve military requirements i.e. formation flying to effect a pairs departure or recovery on the wing in bad weather; low level to stay below radar cover and avoid a range of anti-aircraft weapon systems. Later on, conversion to the Hunter was very quick – by that stage purely flying the aircraft was a given, the imperative was learning to use it to fly in tactical formation, to counter air threats, to carry out air combat against another aircraft, to deliver ordnance, to counter ground threats etc. On the Harrier we had to maintain a wide range of operational skills, even more so with the GR7. Whilst we kept up our basic handling with all the various take-offs and landings that the aircraft could perform and, most importantly with instrument flying, our main efforts were always focused on delivering an operational capability.

The same philosophy applied to our Falcon or B200 operations – a SID, STAR or an ILS approach was a given, the main focus of every flight (apart from transits) was to deliver an effect or an output. Any flying I was going to do in retirement would be back to basics which did not hold any excitement or interest after the flying I had been lucky enough to do over the preceding 40+ years. So whilst I was going to miss the professional challenge and some of the excitement of the Harrier, Falcon and B200 flying, club flying (or even club instructing) was never going to replace it. In a similar vein, although I was by now too old to countenance a move to the airlines, I could look back and say that I was very glad that I had never gone down that route; the money might have been good but I fear I would probably have atrophied through boredom.

In 2013 it was very sadly confirmed that my very good friend Paul Hopkins was suffering from Motor Neurone Disease (MND) and Lynne and I decided that we would carry out some fundraising in support of the MND Association. Part of this involved me contacting everyone who had been on 100 Entry at Cranwell with us in 1969–72 to help raise some money. An old colleague, Wayne Morgan, replied from Spain to say that he had recently arrived out there to work for a large commercial flying school, FTE Jerez. Tongue in cheek, I went back to him and asked if there were any jobs

Paul Hopkins and the author in Spain in 2013, a year before Paul died of Motor Neurone Disease.

going out there? I wasn't really serious as I had not had any thoughts about returning to work. However, he replied straightaway saying that there were a couple of jobs out there that would suit me well. That set me thinking about whether I really would consider going back to work, especially if someone was going to pay for us to go and live in Spain for a few years. Lynne was convinced that it would come to nothing, as had a previous offer to move to Phoenix in Arizona with Oxford Air Training School. Nevertheless, in June, we were invited to visit the school at Jerez de la Frontera and we arranged a three-day visit to look at work, housing and a school for Will who was coming up to his 11th birthday. We were very pleasantly surprised by the school and the environment in Andalucía so when we got home, following a discussion on the practicalities, we decided to give it a go. It would have been very easy to stay in retirement in North Yorkshire, with Will continuing at his present school through to A levels, and we would always wonder whether we shouldn't have gone on a Spanish sabbatical. After weeks of uncertainty over Spanish taxation law and whether we could rent the house we wanted in Jerez, in late August 2014 we were finally good to go. We packed the essentials we would need for a couple of years in Spain into a self-drive hire van, left the house unoccupied, and set off by road for Jerez de la Frontera with Lynne following me in a car with Will, two cocker spaniels and a cat.

Two days before departing for Spain, Paul died at home at Lytham St Annes.

CHAPTER 14
COMMERCIAL FLYING TRAINING: FOUR YEARS IN SPAIN 2014–18

Being mindful of our wish to get Will back into the UK education system for GCSEs, before setting off for Spain we agreed that we would aim to stay in Andalucía for a minimum of two years with the possibility of an extra year if we were all happy; we would only come home early if we were not enjoying our new life in Spain. Since the job was a permanent position, it was up to me how long I stayed. The appointment was titled programme manager which effectively came down to writing and managing the flying programme, although I had deduced from my discussions with Wayne that it would probably come to more than that. In the event, when I got there and started work, my first inclination was to pack up and go home straightaway – in 45 years I had never seen such a dysfunctional flying operation.

Whilst on the surface the flying school was doing a good job producing trained pilots for a number of the world's major airlines, the flying operation was seriously lacking any structured management. There were a number of very well motivated and capable managers but they lacked any organisation that harnessed and focused their efforts. In particular, the chief flying instructor (CFI), Charlie Auty, who was supposed to be managing the operation, was completely rushed off his feet trying to do every job himself. As I said to the CEO at an exec's meeting shortly after my arrival, how Charlie had not had a nervous breakdown was beyond me. Although my job was just to write the flying programme, I could see that the whole operation needed to be sorted out. I was quite alarmed by a lot of what I found but when I asked Wayne what he had done about it before I arrived, he just said nothing really as he knew I would sort it out when I got there. After a couple of months at the school I wrote a first impressions report that spelled out my concerns. By Christmas it had been decided that my role would be changed to head of operations and that I would have responsibility for the whole of the flying operation at FTE; this included the operations dept (despatch of aircraft), the flying instructors (FIs) including their recruitment, aircraft utilisation, the flying programme, and the simulators supporting the flying effort. The CFI would retain responsibility for flying standards and the flying induction of new instructors.

The ex-military base by the airfield provided an excellent commercial flying school site for FTE Jerez with everything the students required on base. This is the view from outside the Flying HQ building where I worked. In the summer the temperature would often exceed 40°C but foggy days like this in the picture were not uncommon in the autumn and winter months.

At this time, the school was training only integrated airline transport pilot licence (ATPL) courses, with some upset recovery training (URT) and renewals of instrument ratings for ex-FTE students who were still trying to get jobs with airlines. The CEO was keen to restart modular training (commercial pilot's licence and instrument rating modules only) and multi-crew pilot's licence courses were under consideration with some airlines. When I arrived at Jerez, the school was operating the single-engine Piper PA-28 for initial training and the Diamond DA42 had recently replaced the Seneca for twin-engine advanced flying including the commercial pilots skills test and the instrument rating. We also had a Slingsby for URT and later on we acquired a Robin in this role as well. Oscar (the CEO) had been keen for me to fly once I was settled into the job; I was sceptical for a number of reasons. Whilst I had a good technical understanding of basic flying instruction from my first RAF tour at Linton-on-Ouse, this was 40 years ago and it was in a military rather than a civilian environment. I had done some civilian club instruction at Staverton in the 1980s but this was very limited and I did not feel that I could bring any true value to the role of FI at FTE. In addition, by this time I had a medical limitation on my licence which restricted me to flying 'as or with co-pilot' and this would restrict quite considerably the flying I could do with ab-initio students. Finally, I knew full well from what I could see around me at FTE that what was required was someone committed full-time to getting the flying operation on a more professional footing. Having me half-focused on going flying was not going to help make this happen.

As I took control of the flying operation, the first thing I did was introduce the novel concept (!) of a weekly managers' meeting to discuss what was going on, keep everyone informed, and to look ahead to the coming week. Since the managers had little if any experience of meetings, when we first met as a management team, I even discussed how to conduct a meeting, what was expected from everyone in preparation and how to participate. With the gradual changeover of managers over the years, the success of these meetings was variable but at least we now had a communication channel amongst the flying management team and everyone knew what was going on. The individual output of the managers in response to actions agreed at meetings was patchy at best; in some cases, this was due to a lack of application but more often it was due to the amount of flying each manager had to do to make up for an ongoing shortage of flight instructors. Whilst I initially relied on walking across the hangar to keep myself updated on engineering matters, having been caught out on a couple of occasions with a failure to communicate between operations and engineering, I also introduced a weekly meeting with the engineers to discuss aircraft availability and forward planning for servicing which worked very well.

My initial focus was on getting the most out of the flying programme; there was not even any organisation as to who was going to write the daily flying programme when I arrived. It appeared to come down to whichever manager was on the ground first (Charlie or Glen Corcoran) and they would make a start on the programme. This ad hoc arrangement had been running for some months before my arrival. The writing of the flying programme, when done properly, was an intricate and lengthy task. Following on from the preparation (i.e. information gathering of the next day's inputs) which went on throughout the day, starting at 1600hrs it was not unusual for the actual writing of the programme to take three to four hours. With the number of students to cater for, the programme was complex; towards the end of my four years at FTE we were programming as many as 100 students per day to fly.

Students were allocated to flying instructors but this again was something of an ad hoc arrangement. I soon discovered that when instructors were absent, for example on holiday, students would be left without anyone to fly with and could go for a week or more without flying. By introducing a more involved allocation of students to instructors, to ensure a student always had an instructor looking after them, we quickly addressed this issue. In time we also started tracking closely which students had been allocated to which instructor in the past so that when a student had to be reallocated to another instructor, if possible he would be reallocated to an instructor with whom he had already established a sound and successful working relationship.

Leave for FIs was another issue that needed to be addressed. The year before my arrival, FIs had largely been granted whatever leave they requested and, unsurprisingly, this had resulted in a lack of instructors at some points in the summer. I was asked to come up with a leave policy that ensured we had the maximum number of FIs available throughout the year. Although the policy I wrote was certainly restrictive and had to be tightly managed by me, once the rationale had been explained to them, the buy-in from the FIs was surprisingly good.

Understandably, Oscar was keen to raise the training output every year although this was not discussed at my level – the first I usually knew about it was when I got sight of what was in store from the business director. This was always a source of some concern as I couldn't deliver the increasing task unless I could recruit enough flight instructors – and there was never any certainty about this – and the engineers were invariably uncertain about their ability to produce enough flying hours with the aircraft. In 2014, the year I arrived, FTE achieved about 22,000 flying training hours. As we moved into 2015, I found that I was being expected to achieve some 2,000 more hours that year without any consideration or discussion as to the feasibility of this plan. An excellent example of Oscar's approach was that late on in 2015, he decided that he wanted to re-commence modular training in January (modular training had not been run at FTE for a few years at this point). Without any discussion that included me, the first I knew was that the marketing department were advertising four course places starting first thing in the new year using the redundant Senecas and I now had to find the instructors to run the course. Fortunately we had two of the 'old and bold' instructors still living locally who, although they were keen to take on the job, no longer had valid instructor ratings or medicals. Somehow, after a mad scramble just before Christmas, we got both these venerable gentlemen requalified ready for January and the course, and subsequent modular courses, went ahead very successfully. Oscar's bullish approach to increasing the training output had been vindicated.

As the years went by and Oscar's demands increased, it was necessary to find ways of getting more and more flying out of the resources. I had always been sceptical of the fact that although the school flew seven days a week, Saturday and Sunday were largely used for solo flying. It was evident that using every day in the week as a normal instructional day would bring about a step change in the flying output. This would mean bringing instructors in to work at the weekend. In addition, with the extended daylight hours in the summer I could see that we could run some sort of split-shift system which would have half the instructors flying in the earlier part of the day (starting as the airfield opened at 0700) and the rest coming to work later and flying, in full daylight, into the evening. Both these changes would have a notable impact on the working practices of the FIs and would need to be discussed and agreed to ensure buy-in and compliance. In the event, when I briefed the FIs on these proposed changes, following the initial scepticism, there was a very good agreement with a number volunteering to work on a Saturday or a Sunday and a happy split between those who wanted to be early birds and those who wanted to do the later shift. I was very happy with this outcome and through these measures, coupled with more effective programming and the new leave policy, we raised the training output by over 36 per cent in my four years at FTE without any change in flying assets apart from the addition of another DA42 simulator and a Robin URT aircraft.

A headache throughout my time in Spain was recruiting and retaining enough FIs to deliver the task. Many of the FIs were quite happy to spend a year or two at Jerez but only as a means of biding their time until an airline would employ them. Whilst they were bonded for their first two years at FTE, the money to be earned in airline employment meant that paying off the bond tended not to be an impediment for them to leave before their two years was up. One of the concerns I had was with the quality of the FIs that I could recruit. Historically, many of the flight instructors at commercial flying schools would have had military backgrounds and, in most of those cases, they would have been trained as QFIs at the RAF Central Flying School in the UK (or the equivalent in other countries). However, with the decline of the military this resource was no longer available. In the civilian world of flying, virtually every pilot who was commercially qualified was employed by an airline; if they weren't, there was probably a reason for this. Either they were straight out of training; they were a late entrant into flying, and therefore had an 'experience versus age' issue; or they weren't very good and no airline would hire them. The net result of this was that in commercial flying, those that couldn't get an airline job tended to end up instructing. That said, some of my younger FIs were actually quite good (as were some of the long-timers) and, mindful of their aspirations, I was always pleased to see the youngsters move on and progress their careers if they had completed their two years at FTE.

Throughout my time in Spain, we saw a steady outflow of FIs to the airlines which was impossible to control. Despite the requirement for a prospective FI at FTE to be prepared to move to another country to live and work, we usually had a continuous stream of CVs arriving at FTE but, for the reasons discussed above, it was not easy to find the wheat amongst the chaff and I certainly didn't get it right on all occasions. It was notable that the better FIs we could recruit came from the UK; generally, the standard of FIs from across the rest of the EU was poor. On occasions, when a candidate showed commitment and motivation, even though they were not up to the entry standard to join us, we would allow them to observe an induction course to see what was required and they would then take part in the next induction. Given the right candidate, this did achieve the aim but it was most concerning that these pilots were EASA[36] qualified flying instructors who were supposedly already teaching people how to fly.

Our HR manager, Rosario 'Charo' Perez, was a charming lady who had been with FTE since it started and had seen it all with FI recruitment. She was a shrewd judge of character and, like all good HR managers, she could be quite ruthless when required. Unfortunately I did have to dismiss three FIs during my time at FTE and on all three occasions I had unstinting support from Charo and from Oscar. As far as selecting FIs was concerned, Charo would do an initial sift of CVs and send me suitable possibilities; I would then whittle those down to those I thought we should interview. Next we invited the FI candidate to come and stay with us on the FTE campus for a few days, attend an interview, carry out a flight check with the CFI and spend some time looking around the town and the local area. If they were successful, they would then go through a three-week induction course before starting to fly with the students. On one occasion, when we were running out of suitable candidates to invite to Spain for interview, I went to the UK for a week and interviewed instructors in London and Leeds. Although this resulted in just four or five of them coming to work at FTE, this was actually quite a good return on investment.

Once settled into my job at FTE I started to get involved in the student recruitment and selection side of the company which was run by the business director, a lovely South African called Frik Schoombee[37]. This took various forms including carrying out the selection of prospective students who were sponsoring themselves to come to the school; the selection of students on behalf of airlines (these would be done either at FTE or back in the UK); and finally, attending various trade fairs, normally in the UK or Ireland, to discuss commercial training with anyone interested in becoming a commercial pilot. I particularly enjoyed the trade fairs where we usually sent a number of staff. The last one I attended epitomised the multi-national nature of the school. When we sat down to eat the night before the event we had two South Africans, an Italian, a Portuguese, a Gibraltarian, a Spaniard and I was the only Brit at the table. When talking to people interested in becoming commercial pilots, I always asked them whether they had considered flying for the military; only if they had thought about this and discounted it should they think seriously about going into a career as a commercial pilot. There were two reasons for this: the military would train you to fly for free compared with a commercial flying school who would charge you over €100,000; and the military flying (and training, although I didn't say this) would be much more interesting and challenging than commercial flying. In terms of job security, I would also advise them to beware the cyclical nature of the airline job market. At this time, in the 45+ years I had been involved in aviation, I had seen a number of marked downturns in the industry, mostly due to oil crises and major conflicts. At the time of writing, since the world has recently been coping with the Covid-19 pandemic and the virtual freeze on commercial flying, this advice was undoubtedly as relevant at the time I was giving it as it has always been in the past.

One issue which was of some interest to me when I went to FTE was the quality of the students I would find. I had hired a number of young civilian pilots whilst I was at Cobham and many of these had been 'self-improvers' who had come up the hard way doing modular training. They had then gone on to flying instruction as a way to get more experience before being hired by an airline or ourselves. The number of FOs we required at Teesside was quite limited and we generally managed to recruit some very capable and well-motivated pilots into these posts. The FTE intake was a somewhat different matter. Whether they had been pre-selected by an airline, or whether they were 'self-sponsored', these students were attending one of the top commercial schools and paying a lot of money for the privilege. Understandably their immediate aim was to get an FO post with one of the major airlines, if possible, before they even graduated. Becoming flight instructors or going into aerial work such as flying with Cobham was definitely not on their radar.

With our military backgrounds, and having had the privilege of the varied careers we had enjoyed, both Wayne and I had trouble seeing the attraction of arriving in the right-hand seat of an airliner straight out of training and spending the rest of your career doing the same type of flying. In the RAF it was quite normal for pilots posted to 'heavy' aircraft to progress to captaincy after two-and-a-half years (Wayne was a case in point on the Vulcan); in the airline world, an ab-initio pilot could take ten years or more on an airline to progress to captaincy. The students were a very mixed bunch, certainly with some 'oddballs' amongst them, and only a limited number would have been realistic contenders for joining the military as pilots. For those that carried out the three flight URT course before leaving FTE, this came as something of a surprise and an eye-opener about a world of flying beyond a standard instrument departure, an ILS approach and a 30-degree angle of bank turn. There were undoubtedly some very capable students at FTE but also a number who only got through the course by the narrowest of margins. Whilst they all had the required academic qualifications and had passed a selection process to come to FTE, one of the major barriers to entry into commercial flying is undoubtedly the cost of training. Fortunately the size of your bank balance, sponsorship by your family or nepotism are not factors with military aircrew selection in the UK. Therefore the military can recruit from a much larger pool of candidates who have acceptable academic qualifications, high aptitude scores and excellent motivation. That said, some of the FTE students had gone to huge efforts to raise the cash to undergo training and there is no doubt that their motivation was outstanding. Unsurprisingly, given that we had about 200 pilots in training at any one time, we had a number of varied personnel issues to deal with. It was of note that the scale of such issues was undoubtedly higher as a percentage amongst the limited number of females in training as opposed to the males. However, that is not to say that some of the female trainees did not prove to be of a very high standard.

Most of the pilot trainees sponsored by airlines were of a sound standard (or better) although there was an interesting cultural issue which came to light with one of the airlines. I was surprised to hear their cadets being told that they had been selected as the 'future captains' of the airline – and this is before they had even started training, passed a single ground exam or gone solo, let alone complete their training at FTE, qualify as an FO with the airline and eventually pass a command course. Unfortunately this led to a number of these students arriving with a certain arrogance which was not appreciated by the other students at FTE. Such an approach was in stark contrast to my experience of training in the military where you were left in no doubt that you had everything to prove before you were going anywhere within the organisation. In the RAF this worked well in keeping us focused and on an even keel (at work if not always socially). In the demanding professional environment of aviation, there was no doubt which approach I preferred. It was of interest that this particular airline provided FTE with a number of trainees who proved to be a challenge and not all made it through to productive service.

The airlines we trained pilots for over this period were mostly European but we also dealt with some airlines outside the EU. I came across one of the more bizarre occurrences of my time at FTE with one airline-sponsored student. The student in question was completing his paperwork one day after a solo flight and was having trouble entering the details into the computer. The duty instructor went to help him and discovered that the take-off and landing times the student was trying to enter did not correlate with the times logged by ATC for his flight. Subsequent investigation by my very astute operations manager, Maria, determined that in virtually all his solo flying, the student had claimed to have flown the time required by the syllabus but had, in fact, flown a lot less time. Having got the facts together, I confronted him with this but he was totally in denial that he had logged incorrect flight times. I suggested he went away and had a think about things and come back to see me the next day. In the morning he agreed that there may have been a bit of difference but he couldn't (or wouldn't) give any explanation for how this had come about. The only logical explanation we could come up with was that he had some fear of flying on his own but he would not own up to this; if he had, this would almost certainly have been the end of his flying career.

Naturally, the next step was to inform the airline about what was going on with this student. Understandably perhaps, they were keen to find a way to resolve the issue and allow him to continue his training. Despite their pleas, and a meeting with Wayne and me at Jerez, we were adamant that the student could not continue his training until the issue was resolved to FTE's satisfaction; not least, the falsification of flight times in aviation for the issue of a licence is a criminal offence. The only lifeline we could offer the airline was to send the student to discuss the issue with an eminent

The sun sets on two of FTE's old Senecas at Jerez de la Frontera and on my career in aviation.

aviation psychologist in London which we did. Unsurprisingly, this interview failed to unlock the impasse between the airline, who still wished to keep the student in training, and FTE, who were refusing to train him any further. Following the interview with the psychologist, the airline had the psychologist's report and the matter lay in their hands; the student was still at FTE awaiting a decision. After some weeks, Oscar had still not received any input from the airline so he instructed that the student was to pack his bags and leave the campus. That was the last we heard of the matter.

The work at FTE had been challenging but I was very pleased with the progress that had been made by flight operations over this period; we had also had a fabulous experience as a family. We finished up staying in Spain for four years, the main reason for our return being Will's schooling. Oscar had been keen for me to remain at FTE for as long as possible and the final year he very kindly agreed that I should move onto a four-day week on the same pay as previously. Given the hours I was working, he was still effectively getting a full week's work out of me in four days! I retired from the company in July 2018 and began my second retirement. After six weeks 'holiday' at home in Spain, we packed up house and returned to the UK.

CHAPTER 15
AN AVIATION CAREER IN RETROSPECT

When I look back on a working life that spanned 49 years, and was for most of the time directly or closely related to aviation, I don't think I could have asked for a better start than to train as a pilot with the Royal Air Force. The discipline that was imposed, and the self-discipline that was required, from day one at Cranwell was a grounding that would shape our lives and play a major part in fitting us well for service as commissioned officers and as pilots. The staff that trained us were of a high standard and their varied backgrounds in the RAF gave us a broad education in what the service was about and what it had to offer. When I consider the military-flying training system that is now in place, I regret that for perhaps understandable reasons of cost-cutting, it is undoubtedly a rather shallow reflection of the training we were privileged to receive.

For the pilots, we all received the same 160-hour wings course; only at that point were we streamed to fast jet, helicopters or 'heavies'. All pilots at this time were trained to a high standard in applied flying to qualify for their wings and could return as QFIs or to other flying roles in the RAF at a later date with the benefit of that training behind them. Not only has the syllabus undergone major changes to ensure students are streamed much earlier, thereby denying many of them such a broad training, but the eclectic mix of QFIs that trained us at Cranwell has now, to a large extent, given way to civilian QFIs at the earlier stages of training[38]. This denies the trainee pilot that military inculcation which had always been a most valuable hallmark of officer and aircrew training in all three services. From my time at FTE of late, I have seen first-hand the quality of the civilian flying instructors being recruited by the military[39] and this gives me serious concern over the efficacy of this arrangement. Perhaps the decline of the military flying training system over the last 50 years can be put down to a classic case of 'knowing the cost of everything and the value of nothing'. One can also look at some of the accidents that are occurring in the civil world – not least the inability to recognise a stalled condition and do something about it – and one can see that sound pilot training these days is arguably in short supply, and I fear that this may become a factor in the UK military if it hasn't already.

In considering a career as a pilot in the military, it is essential to question the ultimate requirement to take life, or to give your own life, in the service of the nation.

To commit to such a contract requires a certain faith in the democratic process in the United Kingdom. Whereas the commission you hold as an officer is from the Crown, the commitment of the armed forces to war fighting or operations is at the behest of the government of the day. I was perhaps fortunate that in my 30 years in the RAF I was never asked to deliver munitions in anger although I did fly in three operational theatres, with a live-armed aircraft under specific rules of engagement. Interestingly, although the primary role of the Harrier was offensive support (i.e. ground attack in the forward area of the land battle), the three operational theatres/ primary roles I deployed to were: Belize/air defence; Falkland Islands/air defence; and Iraq/reconnaissance.

To give your own life was, unfortunately, something that was rather too common in the Harrier Force. Starting my operational career with the Harrier GR3 was mind-focusing. The single-seat low-level role, often in marginal weather, with a navigational system of variable accuracy, very basic flight instruments, potentially lethal VSTOL handling characteristics and limited fuel led to a very high pilot workload. Couple this with a single engine, with huge air intakes that were very effective at swallowing birds, and one had the real potential for a bird strike leading to an engine failure every time you flew. Throughout my time in the RAF I was aware of the risks whenever I flew and, to an extent, this awareness was what helped keep you alive. Despite this, as I have previously covered in this account, there were a number of times when my luck nearly ran out. In retrospect, I am a little surprised that, unlike a number of good friends and colleagues, I did survive my military flying career. In many ways the apparent risks reduced when the Harrier GR5/7 came into service with its excellent navigational suite, hugely improved visibility, additional fuel, much improved flight instruments and very powerful auto-stabilisation in the VSTOL regime. Nevertheless, as discussed previously, there were still four occasions when I was in command of my squadron that I came close to not returning from the mission.

The near-miss with the Tornado in Wales could have happened to any aircraft flying at low level. For example, two of my squadron commander contemporaries, one on Tornado GR1s and the other on Jaguars, had a mid-air collision. All three crews ejected safely. It was the second ejection for the Tornado squadron commander; he had previously ejected from a Harrier GR3. However, the other three incidents (Macrihanish runway, air miss at Yuma and close call with the sea off the Lake District at night) could all be attributed to some extent at least to the level of risk we were prepared to take in our training. And this is the thing: in the 1980s and 1990s (and probably still to this day), the culture of risk-taking on a fast-jet squadron was determined by the squadron commander and his executives. Turning back to my flight commander tour in Germany, I believe we ran a hard and demanding squadron. We knew that the harder we trained in peace, the better we would perform when we went to war (and the greater the chance of survival). The Harrier Force was always blessed with having the better pilots out of the training system and the mindset this engendered was invariably to challenge your capabilities, to seek to improve one's personal performance and the operational capability of the squadron. In Germany, 3(F) Squadron's focus on doing things the hard way (to improve) got us the name of 'Cement Heads', which we wore with pride. Nevertheless, it could, of course, then become difficult to determine whether you had got so caught up in striving to improve your capabilities that you had crossed the invisible line into unsafe operations.

When I was in command of 1(F) Squadron, I was blessed with a very able bunch of flight commanders – most of them went on to be promoted to group captain, 1- or 2-star level and well deserved too – all of whom had the same sense of 'operational improvement'. We now had a much more capable aircraft than the old GR3, but this just meant that we could do more with it, expand our capabilities and challenge ourselves further. This was the mindset that we took into our night trial and a combination of professionalism and good fortune meant we got away with a clean sheet after two winters of teaching ourselves to be night-attack pilots. The issue with night flying was, for me as the squadron commander, to determine what was reasonable to ask the pilots to do and for them to overlay on this their own personal judgement to keep themselves on the safe side of the line. You could not fly the aircraft for them, nor could you remove the risk entirely, but as a team we would always strive to find the right level of risk. Another yardstick of operational capability was the squadron's performance on Operation Warden flying into Northern Iraq. It was gratifying to get unsolicited feedback from one of the intelligence officers in theatre that our approach to the job was more professional and our results noticeably better than the squadron that replaced us.

So how do you create squadron executives who can make the fine judgement in the risk equation? Simply through giving them experience and generating a culture of responsibility and autonomy in your future airborne leaders and in my time the RAF did this extremely well. From my first tour at Linton-on-Ouse when I became a Standards QFI at the age of 23, then just one year into my first tour on the Harrier when I became the squadron QFI, I had great trust placed in me by my seniors. The same responsibility was given to me on the OCU as a display pilot and latterly as the flight commander on B Squadron 233 OCU. But it was when I went to Germany on promotion to squadron leader that I really started to develop as a leader and supervisor of fast-jet operations with command of field sites three times every year, running a training programme for our pilots to deploy to the South Atlantic and then, of course, commanding the Harrier Flight at RAF Stanley on two separate occasions. By the time I assumed command of 1(F) Squadron, even though I had been away from front-line flying for six years, I had a good deal of Harrier flying behind me and an experience of getting the most out of a squadron operationally without, I hoped, compromising safety unnecessarily. Sadly, of course, as mentioned previously and well covered by Bob Marston in *Harrier Boys Volume One* (Grub Street), you didn't have to be in the Harrier Force very long before you realised that you were losing colleagues (and often good friends) on an uncomfortably regular basis. Interestingly, in general (although I can think of the odd exception), this did not dint the striving for operational excellence which was an enduring quality of the RAF Harrier Force throughout its life. When fatalities did occur, the coping mechanism was usually a good party – then get back to business. One of the best and most memorable wakes was in the officers' mess at Wittering for my good friend Bruce Cogram who was killed in a mid-air collision with a Starfighter at low level in Germany in 1985.

But what of the Harrier's operational capability? Certainly in the GR3 era, part of the operational challenge was driven by the limitations of the aircraft, specifically little fuel, relatively low maximum speed and a limited war load of weapons. Our concept of operations did allow for these deficiencies but, if detected at low level by a well-flown and capable fighter (and that was a big if), we had to rely on aggressive low-level manoeuvring and mutual support. With respect to surface-to-air missiles, our only salvation was terrain shielding, flying very low and as fast as possible. And that is why we trained hard. Of course, Tornado GR1 operations in the First Gulf War (which the RAF Harrier did not take part in) showed the fallacy of putting all your training effort into this particular modus operandi. The GR7, and latterly the GR9, addressed many of the limitations of the GR3 and the RAF had by then embraced the concept of prosecuting offensive air operations in the middle airspace with AWACS, effective onboard EW suites and fighter support. Consequently, it was regrettable to see an aircraft that had been so well-developed taken out of service at no notice in 2010, just when it was required off Libya onboard HMS *Ark Royal*, the last RN carrier, which had also just been removed from service. By the time it left service, the RAF Harrier had proved itself in its STOVL[40] capability (the only aircraft that could operate in Belize and in the Falklands War) and in its later days operationally in Iraq, Bosnia, Sierra Leone and Afghanistan. Despite our own perceptions of the Harrier's limitations, especially in the early days, I have no doubt that it was always viewed as a very capable and difficult adversary by those who were on the 'other side of the fence'.

After eight-and-a-half years of Harrier flying, surviving an ejection and a few other close calls, two stints in the South Atlantic running the Harrier Flight and the demands of a very busy tour at Gütersloh, I was ready for a change of scene. My double tour at Barnwood was a very interesting, happy and rewarding period. Besides giving me the chance to adjust to some family stability after the last very hectic three+ years of rushing around with my hair on fire, I quickly learned that a different approach was required for this job. On a fast-jet squadron everything is done at speed, both in the air and on the ground. At Barnwood, looking after people's postings, you had the luxury of taking time to try and make things work for the best and look after your customers. I'm sure I didn't get it right all the time but, as I moved into the wing commander's post, it was gratifying to see some pilots' long-term aspirations that I had set in train coming to fruition. Interestingly, the General Duties (GD) branch of the RAF (the flying branch) was the only branch that was posted by its own officers; all other branches were posted by administrative officers. I was always of the view that posting our own in the GD branch was correct. I am not saying that aircrew were a special case per se, but a good understanding of the different roles and the value of moving (or not moving) pilots to a different appointment would be based on a number of years' working experience in that role by the relevant desk officer. This, I believe, gave the department the ability to acquit itself well. I am sure that some of our customers would not agree with this assessment but a principle one had to accept when moving into Barnwood was that you were never going to satisfy everyone's aspirations. This might be because it was not possible to post a certain officer into a requested appointment because there was no vacancy (or their skill set might be required elsewhere[41]), or it might be that the officer's aspirations were unrealistic. Whilst at Barnwood inter alia I had responsibility for Harrier pilots' postings the whole time I was there. Consequently, barring the six months at JSDC at Greenwich, I was closely connected with the Harrier Force from the time I joined in 1976 to ending my squadron commander's tour in 1994 – a total of 18 years and, not far off half the aircraft's time in service.

Having gained much experience in the postings environment as a squadron leader desk officer, a wing commander (and, for a number of months, acting deputy director covering the group captain post), I would have been very happy at some point in the future to return to the department as director (1-star) however, life was to take a different turn. For understandable reasons, the RAF had a policy of moving officers who were progressing through the rank structure to appointments in a range of different areas to help fit them for higher command. So, even if I had stayed in the RAF and made promotion to air commodore, it was unlikely that I would have returned to personnel management.

As previously discussed, my other staff experience at Logistics Command, my final tour in the RAF, was not a great one. However, it did suggest to me that, at this time, both personnel management and Logistics Command were over-staffed. If this was replicated across other staff areas in the RAF and the other two services, there is little doubt that substantial savings could have been made in the MoD/HQ areas around this time to improve other military capabilities.

Running a front-line squadron was very much a full-on occupation. Whilst I obviously considered carefully my decisions in command at the time, once I had left the squadron I moved on to focus on taking command of RAF Scampton without analysing too much in retrospect what happened. Nevertheless, with the passage of time, and certainly as I came to the end of my working life, I found myself reflecting on what did happen over those memorable and demanding three years in command of No. 1 (Fighter) Squadron. There was, and maybe still is, a view in the RAF that running a front-line squadron could be bad for your career. Certainly, as a fast-jet squadron commander, every day you ran the risk of having a high-profile incident or accident on the squadron for which you could expect to take some or much of the blame. Nevertheless, filling that role would be the pinnacle, in flying terms, for any self-respecting pilot and I was no different in that respect.

During my time as OC 1(F) Squadron, I lost two aircraft: a GR5 due to an engine failure (pilot ejected successfully) and a T4 due to a bird strike[42]. From an aircrew/supervision perspective both these accidents were normal operating hazard accidents and I will not consider them further. In addition, the squadron had a high-profile accident when one of my pilots flew into a huge electricity cable suspended across a fjord in northern Norway; and finally, as previously mentioned, one of my exchange pilots was killed when he flew into the ground whilst flying with No. 233 OCU. Both these accidents were due to pilot error and therefore, not unreasonably, called into question the supervision on the squadron.

The Norway accident happened on a small detachment (four aircraft) and in all honesty, I have to say that the accident was a failure by the section leader who planned to recce the cable but made the mistake of leading his section of three aircraft up the fjord at the same height as the cable. The aim of the detachment was to familiarise pilots with the flying environment in northern Norway (one of the squadron's war deployment options); inter alia this included the varied terrain, the scale of the topography, the light levels at various times of the year, the unpredictable weather and hazards such as masts and cables across fjords which are well known to be difficult to see when flying at low level. The No. 2 hit the cable and was very fortunate to get the aircraft back onto the ground at Tromsø. Whilst it could be argued that the supervision in place on the detachment (the authorising officer) should have identified the error in the plan, the section leader was an experienced Harrier pilot who would have been self-authorising. It would therefore have been quite normal for the authorising officer not to scrutinise in detail the section's plan in their final pre-flight 'out brief'. Whichever way you looked at that accident, the squadron did not cover itself in glory and, ultimately, I was the person responsible even if I was back at Wittering when it happened.

The accident that resulted in the death of my US Marine Corps (USMC) pilot on loan service with the squadron was a tragic event that had a very unfortunate backstory. With the two-season night trial ongoing, we had requested that the USMC exchange officer, who normally served with IV(AC) Squadron, should come to 1(F)

Squadron to bring some experience of night operations from the USMC. This duly took place although, after a short time on the squadron, it became apparent that the pilot who had been sent to join us was having some trouble adjusting to the demands of our operations. NATO exchange officers having trouble adjusting to RAF fast-jet operations was not unusual and unfortunately a number of them were involved in fatal accidents. There were two main reasons for this: the often poor UK weather and our culture of training hard, including a mindset of pressing on in difficult conditions. In this particular case, given our concerns, we allocated him an experienced and trusted pilot as his mentor to fly with him (leading him or flying as his No. 2 since we did not have a Harrier GR5 two-seat variant at this time). He was also only flown in aircraft with a video-recording system which, as we feared, showed up some of his shortcomings. I discussed this with the station commander and it was agreed that the issue would be taken up with our Group HQ and the USMC. The options were that he should be returned to the US permanently or, given that we did not have the capacity to supervise him more closely on the squadron, that he should undergo further training/evaluation by the OCU at Wittering and only return to the squadron if his performance was judged to be satisfactory. It was decided by higher authorities that he should be given a tailor-made two-week course on the OCU but tragically, on his final sortie before returning to the squadron, he flew into the ground in Gloucestershire and was killed. Since he was flying an aircraft with a video-recording system which survived the crash, we were able to see the cockpit view leading up to impact – a harrowing experience. Undoubtedly there was a lot of heart searching after this awful outcome as to whether we had handled the issue correctly. Both at the time and in retrospect, I was of the view that we had been correct in what we had done. These two events aside, my three years in command was, I believe, a happy and successful period for No. 1 which saw the squadron well positioned to move on from operations in Northern Iraq to Bosnia and ultimately, Afghanistan.

A question I ask myself is whether I should have stayed in the RAF and aspired to higher rank? There is no doubt that as an early promotee to squadron leader, wing commander and group captain, selection for the RCDS Course and the Higher Command Staff Course, I was well placed to try and pursue this path. However, there were a number of issues that served to persuade me otherwise. Going back to my early school days, my headmaster wrote on my report when I was eight years old that I was 'a very nice chap but not very bright' and I suppose this comment, consigning me to not having any academic potential, stuck with me throughout my working life. One issue that helped me in my RAF career was that I was very young for my school year; in fact, with a birthday on 12 September, I should have been allocated to the next school year but somehow my parents placed me a year ahead of my peers. I have wondered whether being in the correct school year would not have helped me in terms of academic achievement. Having subsequently performed to a very moderate standard at O and A level, despite having the advantage of attending a very good school, I had no appetite for higher education and was very happy to stay at Cranwell and pursue the fastest route available into flying rather than spend three years at university.

The lack of a degree did not impact my progression through the RAF although I did at times feel intellectually challenged, not in the work place but on staff courses. When I was a wing commander at Barnwood, I gave presentations on the personnel management organisation to various audiences. To lighten the content, I would include some amusing quotes taken from officers' confidential reports. One I used was along the lines of 'I rate this officer's intelligence at 5 [5 out of 9 would be a low assessment] and consequently his promotion ceiling as group captain'; in retrospect I can see this as a reflection of myself[43]. These days people talk about 'impostor syndrome'. Whilst I never felt an impostor at work, I certainly felt this at RCDS and HCSC and this heavily influenced my decision to leave the RAF at the age of 46 rather than stay in the RAF to achieve higher rank. Given my status as a single parent with two teenage sons, the RAF had been very considerate in supporting me in my latter tours of duty. I am aware that I was given command of Scampton as the Air Secretary was adamant that my personal situation should not impact my progression. My later posting to Brampton facilitated my being close to my boys and home at Stamford. Nevertheless, I was not well placed to meet a four to six-month operational posting, of which there were a number for group captains at this time supporting Iraq, Afghanistan and the Balkans. Leaving the RAF would also give me more stability and control over my personal life. My wish to gain a Masters degree (MSc in Air Transport Management at City University, London) whilst I was working for Cobham was, in part, to prove to myself that I was not totally without intelligence and it was gratifying to complete this, at the age of 50, with a Distinction. Nevertheless, there was no doubt in my mind that my fortes, for what they were worth, lay in flying, flying instruction and operations management. My years at Cobham and at FTE built further on my relatively broad military experience and proved to be enjoyable jobs (for the most part) that, like the RAF, had a sense of value and purpose.

My return to flying when I left the RAF gave me a new insight – on the civil register, with a crew and in a completely new role. Whilst we were carrying out what is termed aerial work, as opposed to airline flying, which is commercial air transport, the regulations we operated under related very closely to airline flying (e.g. operating under an air operator's certificate, an operations manual, and being constrained by crew duty flight time limitations etc.). One of the more interesting aspects to our Cobham operation was that we were, in effect, military pilots (flying under the Official Secrets Act) operating under civilian auspices, but most of the time in a military environment. Whereas this did not cause any particular issues with regard to operating to two different sets of regulations (military and civil), it did give rise to some cultural issues relating to the crews who were a mixture of ex-military from a variety of backgrounds and civilian first officers and a few captains[44]. The Cobham Falcons were often being used as quasi military fast jets although they suffered a large number of limitations in this respect (e.g. speed, manoeuvrability, operating ceiling, rate of climb, G limits, visibility from the cockpit, lack of ejection seats etc.). Similar to the Harrier Force, keeping the operation safe was a demanding task, which became more demanding as our capabilities expanded based on our proven ability[45].

I was never in doubt that our output at Teesside for the RAF and allied air forces was good. However, it was interesting that, as far as I am aware, over the ten years I was responsible for delivering Contract 020 to the RAF, the military never carried out any in-depth quality assessment of our output. I think this speaks volumes for the service we provided, that it was never in question even though contracted-out services in the military did not always get a good press. It was quite entertaining to take senior RAF officers like my ex-Harrier colleagues, Clive Loader (at the time Air Marshal Clive Loader Deputy Commander-in-Chief HQ Strike Command) and David Walker (at the time Air Vice-Marshal David Walker Air Officer Commanding HQ 1 Group), flying in the Falcon to see their surprise at what we were doing with the aircraft and the profiles we were flying.

I have little to add to the comments I have made about the management challenges when I was at Cobham. On the Falcon side, notwithstanding some of the difficulties, the Falcon operation at Teesside invariably produced a quality product, a high number of flying hours every year with an efficient use of crew. On the other side of the hangar at Teesside, whilst operating with FPL was nearly always a good flying experience with a great bunch of guys, I felt the business unit (BU) was badly let down by a top management team at Bournemouth that failed to give calibration the focus it needed. Latterly, efforts were made to rescue the situation. Firstly, their BU manager wanted FPL to have their own air operator's certificate instead of operating under the Cobham AOC. I could never see the point of this due to the expense of setting this up and, more importantly, the additional management overhead this would entail. Sure enough, once they had their own AOC and I relinquished my role of operations director, I would see their three senior pilots (the chief pilot and two trainers) spending most of their time in the office and doing a very limited amount of flying. Secondly, getting rid of the incumbent BU manager and replacing him with a manager who was based at Bournemouth (adding further to the remoteness of the top management). Thirdly, sometime after I had left Cobham, a decision was taken to move the BU from Teesside to Bournemouth. There were probably good reasons for this but there were also strong arguments for not doing it (e.g. loss of key staff who would not move, cost of moving staff to the Bournemouth area, increased transit times to many UK airfields for calibrating etc.). Nevertheless, not long after the move to Bournemouth, the BU was closed down and Cobham left the calibration business.

The top management at Bournemouth was an interesting evolution during my time at Teesside. The MD who was in place when I joined Cobham was replaced by Alex Hannam not long after I took over from JD. Alex was a hard-hitter who was brought in to tighten things up which he did very successfully and turned the special missions operation around in the space of a few years. Whilst I, amongst others, never felt relaxed in his presence, there was no doubt that his no-nonsense attitude and disciplined approach brought great benefit to the company and to those of us working there[46]. But Alex wasn't just about saving cost; he also invested in areas where money needed to be spent and we benefitted from this at Teesside where our working accommodation was all upgraded to a very high standard. After Alex left Bournemouth to move onto the main Cobham Board, we had various other MDs who took advantage of his legacy but never seemed to have his firm grip. Unfortunately this covered the difficult period for FPL which included the lightweight-calibration disaster. Eventually, shortly before I retired from Cobham, a new MD was brought in from Cobham Australia who came in with some of his team from Down Under[47]. They presided over the failure of the flight calibration BU and ultimately, very sadly, Cobham *in toto* was sold to the US in 2019[48].

The job I finished up in at FTE in 2014 drew on much of the experience I had gained throughout my career up to that point: basic flying training; civilian flight operations; risk management; manpower planning; resource management; personnel management; recruitment and selection. One of the issues we faced when it was decided that I would take on the new head of operations role at FTE was that this appointment did not fit into the flight operations management structure laid down in our operations manual under AESA (the Spanish aviation safety and security agency). Under their structure the CFI was supposed to be responsible for the full breadth of the operation. Charlie Auty was a very able flying instructor and examiner, but he had no training or experience as a manager and I believe this situation is probably typical of many commercial flying training schools today. They no longer have the ex-military flying instructors available who also had some experience of organising and managing flight operations. Oscar was very pleased to have Wayne and me sorting out the flying operation; he said that it was the first time since he had taken over the company some years previously that he was confident that the department was being run competently. FTE had gone through a surprising number of heads of training and CFIs during Oscar's time as CEO. As regards my appointment not fitting into the operations manual management structure, we chose to ignore this and AESA never commented on it.

As previously discussed, there is undoubtedly a marked difference between the quality of instructors operating at a commercial flying school compared with the QFIs I worked with during my RAF career. But, in mitigation, the output from a commercial flying school is limited to a CPL skills test and an instrument rating. Whilst these tests are certainly demanding, the military pilot that we were training in the 1970s needed to cover these areas (in some respects to a higher standard) and also be proficient at low-level navigation (250 ft), aerobatics and formation, on a more demanding aircraft in about the same number of training hours as his civilian counterpart.

Concern has been expressed in various forums about the training standards and examining in the commercial aviation environment. During my time in Spain, I had contact with a number of the industry regulators across Europe, all of whom were working collectively under the European Aviation Safety Agency (EASA). EASA was formed to create a common standard of best practice in aviation across Europe. Unfortunately, whilst the common standard was being met on paper, in practice this was not always the case. As I left the industry in 2018, I had the following concerns (without attribution to the specific authorities concerned):

- Flight operations inspectors who were not sufficiently skilled to carry out the role, or to operate to the level of the ratings on their licence, coupled with an authority that was aware of, but failing to address the issue.
- Inadequate testing of students for the issue of a commercial flight crew licence (not at FTE).
- EASA-qualified flying instructors who were not capable of imparting knowledge either in the air or on the ground.
- A culture of only inviting examiners back to examine students at a commercial flying school if they always passed students on flight tests (not at FTE).
- An inability by an authority to audit a flight training organisation effectively.
- No psychological testing of prospective pilots to determine their suitability for the role of an airline pilot.

At the level of basic commercial flying training, a combination of some poor-quality students, instructors of variable ability and a regulatory system that had a number of failings could lead to qualified airline pilots who are not fit for purpose. I believe that there are two factors which, thankfully, protect us from this outcome: the training and regulation at the airline level; and the increasing automation on the airliners of the 21st century. By virtue of having been selected to join an airline, gaining promotion to captaincy and then being selected for training as an instructor and/or examiner, instructors/examiners on airlines tend to be experienced and capable. They are the people that are required to take the output from the flying training schools and train them to be competent first officers. This system worked perfectly well within Cobham and I believe that it is generally working satisfactorily across the airlines. As regards automation on airliners, despite the recent catalogue of disasters surrounding the Boeing 737 MAX, the flight systems onboard modern airliners in service mean that most pilots are very rarely required to exercise any great skill in actually flying the aircraft. On the odd occasion when they are, unfortunately this can show up their lack of effective training. However, the efficacy of these flight systems in maintaining safe flight in difficult conditions, means that the safety of flying commercially has shown a marked improvement over the last 11 years, notwithstanding a perceived reduction in pilot-training standards.

Without making a claim to having any particular expertise in this area, I wish to add a few comments on the psychological aspects of flying as a professional pilot. I have long been interested in the qualities needed in a pilot relating to the taking of risks and how these might vary between a military and a civilian pilot. In my time in the RAF before going into my first ground tour at Barnwood, I believe I had developed a strong sense of balancing risks. Whilst this was not particular to flying the Harrier or fast jets, the single-seat role and the challenges involved meant that every mission you flew was a continuous risk assessment process that you, and only you, had the responsibility for undertaking. But why would we willingly, and happily, fly around at 250 ft or 100 ft at 420–500 knots, in many respects 'flying by the seat of our pants', when the slightest inattention or miscalculation could quite easily lead to flying into the ground and certain death? Or, at best, we might just hit an obstruction and, if lucky, we might get the chance to eject. The simple answer was that it was a challenge and it was undeniably exciting. The ability to plan a complex mission and take a formation of Harriers through demanding terrain and weather, avoiding or bettering air-defence fighters and SAMs, was professionally hugely satisfying. So how did the RAF go about finding people that were prepared to do these things and yet they were also expected not to have accidents (and were held severely responsible when things did go wrong of their own making)? In effect, they were looking for risk-takers who were somehow going to get it right most of the time in peacetime. I suppose the answer in my day was to find the right combination of youthful exuberance and aptitude and let inculcation play its part, building on the legacy of World War II and the early jet era. It was absolutely the case that all the Harrier pilots I operated with were very willing to take on such risks and back their own judgement in staying the right side of the line. Despite this confidence, it didn't always work out and a number of them didn't live to write their stories. And yet we continued to do what we did.

On reading a draft of this book, my wife commented on the fact that every day we flew we must have been living with the real possibility of an accident, and perhaps never coming home, and that this would have been ever-present in our consciousness. For myself I know that this was so but, in my experience, it was never anything we discussed amongst ourselves. That said, I am aware of one trainee Harrier pilot at Wittering who did express his understandable concern over his chances of survival if the engine failed in jet-borne flight or in the hover. He was taken off the Harrier, moved onto another fast jet and went on to have a very successful career in the RAF. I think the rest of us just internalised our concerns and lived with the risk. There is no doubt that every time you strapped into an ejection seat you were from that moment on about one second away from the need to pull the yellow and black handle. This even applied on the ground, never more so than at RAF Stanley on my first detachment there when our aircraft were parked right next to the runway. With Phantoms landing and taking the hook wire very close to us, if anything had gone wrong with one of these aircraft losing control and departing the runway towards our aircraft we would have had to eject to save our lives. For this reason, we would take the pins out of the seat as soon as we were strapped in and I would always watch any landings that were taking place in case I needed to depart the scene in a hurry. It was fortunate that the GR3 had zero-zero ejection seats (i.e. they could be operated at ground level with no forward speed). Of course, the risk inherent in military flying is not confined to fast jets. I can also recall the loss of a Vulcan, a Victor, a Shackleton, helicopters, Nimrods and Hercules amongst other aircraft during my career. Nevertheless, there is no doubt that the loss rate on fast jets was much higher and the Harrier was on the top of that tree. My course on the OCU at Wittering was quite fortunate: between the seven of us we clocked up a large number of tours on the Harrier with just three ejections and no fatalities. With due acknowledgment to NCO aircrew and the RAF Regiment, there is an amusing adage that RAF airmen have it right – they send their officers off to do the fighting. Yet it is notable that during its time in service, the RAF did not lose any Harrier pilots in the Falklands Conflict, in Northern Iraq, in the Balkans, or in Afghanistan – all our fatalities were in peacetime accidents.

So how does the risk in military flying compare with commercial flying? Obviously there are also many risks involved in airline operations but these can be largely mitigated through rules and procedures. The majority of airline flying is also conducted in heavily protected airspace where deconfliction from other traffic is effected by air traffic control and the avoidance of terrain is allowed for in the structure of the airspace. This removes two of the major risk factors which are inherent in military flying. If you can write enough rules and procedures to cover all perceived eventualities, then your airline pilot just needs to follow those rules and procedures and he and his aircraft will probably stay safe. And, apart from major technical failures or a lack of skill on the part of the pilot, this works very well. But how do you take the adrenalin junkie fast-jet pilot and turn him into a rule-bound airline pilot? I think the answer to this is that even most fighter pilots grow up eventually and, being happy to have survived their time in the military, are prepared to abide by the regulations to ensure their survival (and that of their passengers).

An issue that came to a head when I was at FTE was the psychological suitability of pilots to fly an airliner; this was largely a result of the tragic Germanwings incident[49]. I have previously covered the case of a student who appeared to have some fear of flying solo and yet was keen to progress with his training to become an airline pilot. I also dealt with a case at FTE where another student developed a total fear of flying. In this case he was open and honest about it and, despite our best efforts to help him – although we had little confidence that his situation was recoverable – that was the end of his flying career. Germanwings drew into stark relief the fact that at FTE (and, I'm sure, all other commercial flying schools) no-one was assessing the psychological suitability of our self-sponsored students in any structured way for a career as airline pilots. We were also not aware of what assessments were being made of the pilots we were training for sponsoring airlines. Accepting that this was a very difficult issue to legislate, whilst the Germanwings crash occurred in early 2015, when I left Spain almost three-and-a-half years later, the regulations to cover the assessment of pilot psychological suitability had still not been laid down in law.

Finally, when is the right time to leave the flying to others? Due to the demands of quick reactions and mental agility under extreme pressure, fast-jet flying tends to be the preserve of the younger generation in their 20s and 30s, although squadron commanders, like me, could still be at the forefront of flying operations in their 40s. It may be stating the obvious, but I believe the optimum balance between flying experience and age was the flight commanders in their 30s. They usually had the experience of two to three tours of fast-jet flying behind them, the confidence and ability to lead from the front and the judgement to keep the operation safe. I did a lot of demanding flying on my flight commander tour in Germany and in the Falklands; once settled into the operation at Gütersloh I do not recall having any serious self-doubt over my abilities in the role. Few pilots continued flying fast jets into their 40s and beyond. Unfortunately, a number who did, especially those in command appointments, made mistakes which turned out to be fatal.

One of the earliest station commanders at Wittering in the Harrier era was killed flying the aircraft, as was my good friend Dave Haward who was just taking command of the station in 1998. I have previously covered the very sad cases of Keith Holland, who died on a field site in Germany as OC IV(AC) Squadron, and Bill Green who died in a Tornado accident at night. A contemporary of mine, Wing Commander Nick Slater was killed in 1995 when he was commanding the Strike Attack Operational Evaluation Unit at Boscombe Down. He flew into the Solway Firth in daylight, almost certainly distracted by recording the task he was undertaking with his head in the cockpit and not looking out. One of my elderly fellow instructors on the OCU at Wittering crashed a Harrier GR3 through mishandling the aircraft in the VSTOL regime and was very lucky to survive. Two contemporaries of mine who were commanding squadrons (Tornado and

Jaguar) were involved in a completely random mid-air collision. The Jaguar squadron commander and the crew of the Tornado (the squadron commander and his navigator) all ejected safely and were picked up by the same RAF Search and Rescue helicopter. Having not seen the other aircraft, neither crew had any idea what had happened and only started to work it out when they found themselves in the same rescue helicopter.

Air Marshal Sir Ken Hayr, a very capable and widely respected fighter pilot and a previous OC 1(F) Squadron, was killed at an air show at Biggin Hill in 2001 flying a Vampire when he was 66 years old. Air Chief Marshal Sir John Allison crashed an Me 109 at a Duxford display in 1997 and was very fortunate to survive the accident. I also recall an occasion when the inspector of Flight Safety (a 1-star ex-Harrier pilot) returned from a solo Harrier flight at Wittering. I was duty authorising officer in operations and was asked by the engineers to go out to the aircraft which the air commodore had just vacated as it was unsafe. When I got there, I found that, besides leaving the batteries switched on, he had failed to make the ejection seat safe before climbing out. This could have led to him firing the ejection seat on vacating the cockpit (it had happened before). I spoke with him afterwards and I suspect he was preoccupied with so many work issues that his brain had just switched off from flying as soon as he had landed, a dangerous thing to do, especially in a single-seat fast jet. Whilst I was delighted to be given command of a squadron, I was very conscious of the fact that, by that time, I had been away from front-line flying for six years. It was understandable that the RAF wished to capitalise on my personnel management experience by re-touring me at Barnwood. However, there would also be a very strong argument, in the face of this catalogue of errors by senior/relatively elderly military pilots, that the flying currency and abilities of senior officers could have been taken even more into account before posting them to command flying appointments.

Given that the demands made on airline pilots are quite different in many respects to those made on fast-jet pilots, there is no reason why airline pilots cannot continue flying perfectly safely until much later in life. Present experience would suggest that the current general retirement age of 65 is reasonable for commercial pilots. But there is no doubt that at some point age takes its toll. In 2012 a very experienced fighter pilot I knew (Hunters and Jaguars), and latterly a civil airline captain who was also a keen light-aircraft pilot and instructor, inadvertently walked into the propeller of his aircraft and was killed; he was 67 years old. Everyone has to make their own decision as to when is the right time to stop flying. For me it was easy. By the time I was 60 I had done a lot of different flying and, after a certain point, you don't get any better as you get older. I decided to quit while I was ahead with plenty of other things I wanted to do in life. Anyway, by that time I had almost certainly used up all of my nine lives.

APPENDIX
POWERED AIRCRAFT FLOWN BY TYPE 1969–2011

Below is a list in chronological order of the aircraft flown by the author. (C) for civilian, (M) for Military

DH Chipmunk 1 (C)
Beagle Pup 150 (C)
Cessna 150 (C)
Jet Provost T Mk. 3 (M)
Jet Provost T Mk. 4 (M)
Jet Provost T Mk. 5 (M)
Gnat T1 (M)
Wessex* (M)
Jet Provost T Mk. 3A (M)
Jet Provost T Mk. 5A (M)
Chipmunk T10 (M)
Hunter T7* (M)
Hunter F6 (M)
Hunter FGA9 (M)
Whirlwind* (M)
Gazelle* (M)
Harrier T4 (M)
Harrier GR3 (M)
Meteor T7* (M)
Lightning T5* (M)
Jaguar T2* (M)
PA-38 Tomahawk (C)
PA-28 Warrior (C)
Hawk T1A (M)
Harrier GR5 (M)
Harrier GR5A (M)
Harrier GR7 (M)
Tucano (M)
Bulldog (M)
Jetstream* (M)
Firefly* (M)
Vigilant (M)
Spitfire IX* (C)
PA-23 Aztec (C)
Falcon 20 (C)
Tutor (M)
Beech King Air 200 (C)

* denotes not as Pilot in Command

GLOSSARY

ACMI – Air Combat Manoeuvring Instrumentation

AESA – Spanish Aviation Safety and Security Agency

APC – Armoured Personnel Carrier

APU – Auxiliary Power Unit

ATC – Air Traffic Control

ATPL – Airline Transport Pilot Licence

BALPA – British Airline Pilots Association Union

CFI – Chief Flying Instructor

CFS – Central Flying School

COC – Combined Operations Centre

DFO – Director of Flight Operations

EASA – European Aviation Safety Agency

EO – Electro Optical

EW – Electronic Warfare

EWO – Electronic Warfare Officer

FEZ – Fighter Engagement Zone

FLIR – Forward-Looking Infrared

FO – First Officer

FOD – Foreign Object Damage

FRA – First Run Attack

GCA – Ground-Controlled Approach

HCSC – Higher Command Staff Course

ILS – Instrument Landing System

JACIG – Joint Arms Control Inspection Group

JMC – Joint Maritime Courses

JSDC – Joint Services Defence College

NBC – Nuclear, Biological and Chemical

NVG – Night-Vision Goggles

OCU – Operational Conversion Unit

OPEVAL – Operational Evaluation

PPL – Private Pilot's Licence

PVR – Premature Voluntary Release

QFI – Qualified Flying Instructor

QHI – Qualified Helicopter Instructor

Radalt – Radio Altimeter

RCDS – Royal College of Defence Studies

SAM – Surface-to-Air Missile

SAOEU – Strike Attack Operational Evaluation Unit

SAP – Simulated Attack Profile

SEngO – Senior Engineering Officer

SID – Standard Instrument Departure

SFO – Senior First Officer

SNCO – Senior Non-Commissioned Officer

STAR – Standard Arrival Procedure

STOVL – Short Take-off and Vertical Landing

TMTS – Trade Management Training School

TOT – Time on Target

VSTOL – Vertical Short Take-off and Landing

VTO – Vertical Take-off

ENDNOTES

CHAPTER 1

1 Ground loop is the result of a loss of directional control on the ground, usually on landing, when the aircraft swings violently out of control. Aircraft with a nosewheel configuration do not suffer from this tendency.

2 The technique of selecting and holding the required visual attitude and using the trim wheel to relieve the load on the control column.

CHAPTER 2

3 AVM F. D. Hughes had an outstanding record as a World War II night-fighter pilot being awarded a DSO and three DFCs; later in his career he was awarded the AFC and made a CBE and CB.

4 STUPRECC: Speed, Trim, Unlock, Power Off, Raise the guard on the standby trim, Exhaust the accumulator, Check the trim, Changeover.

CHAPTER 4

5 I discovered later on in my career that this was a case of 'risky shift'. Outside of flying, this is a psychological phenomenon that refers to decisions made by groups being riskier than decisions made by individuals. In a flying context, this term is used to describe the situation where the pilot who is supposed to be making the decisions is overruled by another pilot who may be operating in a supervisory role (in this case my No. 2 who was a senior officer and a very experienced fighter pilot and Harrier operator). Once the decision making has effectively been subsumed by the other pilot, the original decision maker may then decide to leave the decision making to the other pilot. This can then lead to a situation where neither pilot is clear about who is making the decisions and, consequently, dangerous situations can develop. Risky shift is one reason why flying two captains together in an aircraft is not ideal. Captain/first officer provides a 'cockpit gradient' which is explicit that decision making is clearly vested in the left-hand seat. In a military fast-jet context, it is essential that more senior officers who are not acting as the flight lead only intervene in flight when absolutely necessary (usually on safety grounds) and then quickly revert to their subservient role as a member of the formation; in my experience, this generally was the case.

CHAPTER 5

6 One of the forward air controllers we did a lot of work with was Flight Lieutenant Garth Hawkins, sadly killed in a Sea King crash during the Falklands War in 1982 with a number of SAS troops as they were being inserted onto the islands. Garth was the choice of FACs for the Harriers operating in theatre.

7 This amusing occurrence may have been with 3(F) Squadron on my detachment to them for a field deployment the following year.

CHAPTER 6

8 Deflecting the nozzles (i.e. the engine thrust) in normal flight as opposed to take-off or landing.

9 In my experience it was not generally discussed amongst us but it was an inescapable fact that a major control problem, the loss of control, or an engine failure in the hover or at very slow speed would invariably lead to serious injury (if you survived the crash) or, more likely, the loss of your life; the time to react to the problem would not leave sufficient time to eject within safe parameters. Brian Weatherley died when he lost control on a display in Germany, ejecting outside the seat performance envelope. Tim Ellison was paralysed from the waist down when the engine failed in the hover and he was still in the

cockpit when the aircraft impacted the ground. Geoff Timms survived when he lost control carrying out a mini-circuit (jet-borne manoeuvring) and was still in the cockpit on impact; fortunately the aircraft impacted wings level which made the accident survivable.

10 A good friend and an outstanding officer who was very sadly killed flying at night on a work-up sortie for the First Gulf War whilst in command of a Tornado squadron. Bill had just been promoted to group captain and was due to leave the squadron shortly. He would surely have gone on to bigger and better things but for this tragic accident.

CHAPTER 7

11 Tragically, Wing Commander Keith Holland, one of the most entertaining and charismatic Harrier pilots was caught out by this trap. As OC IV(AC) Squadron, in the post-Falklands War period, the RAF Germany Harrier Force were still exercising in the field as normal although a number of pilots from Germany had been used to reinforce 1(F) Squadron in the South Atlantic. As his squadron was about to be split to the four winds to support the effort in the Falklands, Keith decided to do a tour of his squadron's field sites to visit all the squadron personnel. Most unusually, he took a T4 from base which, as usual for on-base aircraft, was equipped with 100-gallon drop tanks. When he came to depart one of the field sites, the ground crew had filled the aircraft full of fuel and Keith failed to spot the error when he checked the F700. On take-off, besides trying to depart in an overweight aircraft, Keith failed to lower flap which was essential; consequently the aircraft failed to clear the trees at the end of the strip and Keith was killed on impact. Such errors of omission were a known hazard caused by distraction and were a real risk, especially for single-seat pilots in command positions.

12 Peter Squire (later Chief of the Air Staff Air Chief Marshal Sir Peter Squire) provides a surprisingly frank account of the interaction between the carrier's senior management and his squadron in *Harrier Boys Volume One* by Bob Marston.

13 The runway we were using at RAF Stanley was AM2 matting, a system of interlocking steel planks which the US had supplied to the MoD. This was laid on top of the original runway which had been subject to damage during the war including a 1,000-lb bomb from an RAF Vulcan Black Buck mission.

14 Back in the UK air-to-air refuelling was normally conducted at medium/high level (20,000 ft or more), but in the Falklands we would routinely refuel at low level (2,000 ft or even 1,500 ft). The main reason for this was that we would often plan our refuelling around a time when we were operating with the C-130 at low level and it would waste a lot of time for them to climb up to medium level. The downsides to this were: firstly, that it could be turbulent at low level making the basket move around a lot which could make refuelling difficult; and secondly, if you made a gross mishandling error, it would be possible to flame out our single engine which would lead to an immediate ejection at that height. (The F-4s also refuelled at low level with us but they had two engines to obviate this particular risk.) Given that the C-130's refuelling basket was large and we were confident in our own abilities at refuelling, we didn't consider these issues to be a major concern.

15 Some years later, now OC 1(F) Squadron, Leaks ejected from a Harrier GR7 into the Mediterranean. Since it was his second ejection, he was given a full body scan which revealed an inoperable tumour on the brain. That was the end of his flying career although he remained in the RAF achieving the rank of air commodore. His son, Joshua, joined the Parachute Regiment and was awarded the VC for his gallant actions whilst on operations in Afghanistan.

16 The outgoing boss, Bob Holliday [one of the flight commanders when I had been on 1(F) Squadron], was a good guy to work for with a very relaxed style. Just before he left Gütersloh, he was driving through the married quarters area as I was walking down the road. He stopped, motored down the window and said, "I've just heard I've been awarded the OBE – Other Buggers' Efforts – absolutely right. Thank you very much!" He then wound up the window and drove off.

CHAPTER 8

17 The first pilot exchange boarding I did at Barnwood was for an RF4 post in the US and I included a single officer who was my preferred choice; this was agreed by my wing commander and group captain.

However, the British embassy in Washington kicked back at the concept of accepting a single officer on exchange but my candidate was supported by the department and the officer went to the States as planned. He returned three years later with an American wife.

CHAPTER 9

18 After landing, the nosewheel T-handle was pulled to open the nosewheel doors to allow a visual inspection of the nosewheel bay. The T-handle had to be pushed in prior to take-off otherwise the undercarriage would not retract.

19 Nightrider was the name given to the Harrier T4 modified for night-trial flying.

20 A system to blow the night vision goggles off the pilot's helmet in the event of an ejection. If this did not take place, the weight distribution with the goggles still attached to the helmet would probably result in the pilot's neck being broken on ejection.

21 Forward looking infrared. The combination of NVG and FLIR is often referred to electro-optical (EO) operations.

22 Green lit panels which provided a visual reference for close formation flying at night. They were green to avoid compromising the NVGs.

23 In comparison, a full moon might provide 90 millilux or more. In such conditions, in dry air, the performance of the goggles was outstanding although never anywhere near daylight visibility.

24 A 'swept' position on the leader with the range being adjusted as required depending on the weather, terrain and manoeuvre.

25 Cloud cover is measured in 'octas' or eights of cover e.g. five octas = 5/8ths.

26 An integrated global positioning system/inertial navigation system. This provided a very accurate GPS position which ensured that the moving map display in the cockpit driven by the INS was always reliable.

27 A key pad mounted directly in front of the pilot's face which is used to enter navigational data, and for many other functions, both before and during flight.

28 The 'cultural lighting' from runway lights is so bright that it would cause the goggles to be swamped out by an excess of light and become unusable.

CHAPTER 10

29 Nigger was the black Labrador owned by Wing Commander Guy Gibson, Officer Commanding 617 The Dam Busters Squadron. He was run over and killed just outside the station the day before the Dams Raid and was buried outside Guy Gibson's office whilst the raid was in progress on his owner's instructions.

CHAPTER 11

30 I spent 14 months filling a wing commander appointment, despite being ranked as a group captain, before taking command of RAF Scampton; I then spent 16 months in higher staff training.

CHAPTER 12

31 The intention had been to purchase the aircraft. I was informed that the purchase was not approved by the Cobham Board but a commitment had already been made to the manufacturer. As a result, the manufacturer agreed to Cobham backing out of the purchase on the condition that a lease was taken on the aircraft.

32 That said, it was not always possible to recruit the preferred candidate at the right time since officers leaving the military are on a defined timeline which did not always meet our needs.

33 Such events were not unknown. In 1982, an RAF Phantom shot down an RAF Jaguar in Germany with an AIM-9G Sidewinder air-to-air missile. This was the culmination of a series of organisational and operating errors. Fortunately the pilot, who I knew, lived to tell the tale. Had it been an AIM-9L missile he probably wouldn't have survived.

34 Air Traffic Information Service which continuously transmits the arrival airfield's weather, runway in use and any other relevant information.

CHAPTER 13

35 The origins of KOS can be traced back to the friendship and collaboration between Dave Harle and myself in the resurrection of the Cranwell Potholing Club in 1969/70. Amongst others, Paul Hopkins also took part in some of these caving trips. Dave and I started backpacking in the mid-late 1970s, which was when KOS really started. Paul then joined in our walking exploits, but only once we had graduated to more comfortable accommodation, followed in the late 1980s by John Danning, so the initial link for membership was 'RAF pilots'. More recently we have, very magnanimously, embraced members without a pilot background.

CHAPTER 14

36 European Aviation and Safety Agency responsible for aviation standards and safety across the EU.

37 Frik had a very interesting background as a civil servant in South Africa working closely for both de Klerk and then Mandela during the apartheid transition period.

CHAPTER 15

38 Military Elementary Flying Training and Flying Selection has employed civilian flying instructors for many years but originally these were all ex-military QFIs trained at CFS, normally A2 or A1 qualified.

39 Whilst at Teesside I did recruit two young flying instructors who had been instructing in the military training system and they were very competent, possibly because the military had trained them well. They were certainly a cut above the instructors who left FTE to join the MFTS as QFIs.

40 Although the Harrier was often referred to as a VSTOL aircraft (Vertical Short Take-off and Landing), due to its very limited vertical take-off performance, operationally the aircraft was more correctly termed STOVL (Short Take-off and Vertical Landing).

41 My posting to the OCU at Wittering and my inability to extricate myself by volunteering for other appointments being a case in point.

42 This was Ashley Stevenson's second ejection inside 12 months. His account of this ejection is in *Harrier Boys Volume One*.

43 One of my favourite quotes was: "There is only one person on this station who drinks as much as this officer and that is his wife."

44 My MSc thesis 'A Safety Culture for Civil Air Operations in a Military Environment' examined this issue.

45 Unfortunately the risks involved in the Falcon operation were graphically illustrated by a mid-air collision between a Learjet (undertaking a similar role to the Falcon) and a German Luftwaffe Typhoon in 2014. Both the ex-Luftwaffe pilots in the Learjet were killed; the Typhoon made a successful emergency landing.

46 A sad reflection on the mentality of some of my aircrew at Teesside was that when Alex announced an all-employee Christmas bonus of £500 there were two complaints: that it was taxable and that it wasn't linked to pay. There was no way of pleasing some people.

47 In my ten years running the Teesside operation I worked under six different MDs at Bournemouth.

48 It should be pointed out that Cobham was a very large company, at times getting into the FTSE100. The Flight Operations Division in which I worked was only a small part of Cobham and the only part that operated aircraft.

49. On 24 March 2015, the first officer of Germanwings flight 9525 locked the cabin door when the captain briefly left the flight deck and flew the aircraft into the Alps; all aboard were killed.

INDEX